THE DOMINANCE OF MANAGEMENT

Voices in Development Management

Series Editor:
Margaret Grieco
Napier University, Scotland

The Voices in Development Management series provides a forum in which grass roots organisations and development practitioners can voice their views and present their perspectives along with the conventional development experts. Many of the volumes in the series will contain explicit debates between various voices in development and permit the suite of neglected development issues such as gender and transport or the microcredit needs of low income communities to receive appropriate public and professional attention.

Also in the series

Losing Paradise
The Water Crisis in the Mediterranean
Edited by Gail Holst-Warhaft and Tammo Steenhuis
ISBN 978 0 7546 7573 0

Tourism, Development and Terrorism in Bali
Michael Hitchcock and I Nyoman Darma Putra
ISBN 978 0 7546 4866 6

Women Miners in Developing Countries
Pit Women and Others
Edited by Kuntala Lahiri-Dutt and Martha Macintyre
ISBN 978 0 7546 4650 1

Africa's Development in the Twenty-first Century
Pertinent Socio-Economic and Development Issues
Edited by Kwadwo Konadu-Agyemang and Kwamina Panford
ISBN 978 0 7546 4478 1

Social Exclusion and the Remaking of Social Networks
Robert Strathdee
ISBN 978 0 7546 3815 5

The Dominance of Management

A Participatory Critique

LEONARD HOLMES
Roehampton University, UK

LONDON AND NEW YORK

First published 2010 by Ashgate Publishing

2 Park Square, Milton Park, Abingdon, Oxon OX14 4RN
711 Third Avenue, New York, NY 10017, USA

Routledge is an imprint of the Taylor & Francis Group, an informa business

First issued in paperback 2016

British Library Cataloguing in Publication Data
Holmes, Leonard.
The dominance of management : a participatory critique. --
(Voices in development management)
1. Community development, Urban--Management. 2. Community development, Urban--Citizen participation. 3. Community development, Urban--England--Case studies. 4. Community development--Developing countries.
I. Title II. Series
307.1'416-dc22

Library of Congress Cataloging-in-Publication Data
Holmes, Leonard.
The dominance of management : a participatory critique / by Leonard Holmes.
p. cm. -- (Voices in development management)
Includes bibliographical references and index.
ISBN 978-0-7546-1184-4 (hardback) -- ISBN 978-1-4094-1047-8 (ebook)
1. Management--Research--Great Britain. 2. Management--Great Britain--London--Case studies. 3. Community development--Great Britain--London--Case studies. I. Title.
HD30.42.G7.H65 2010
302.3'5--dc22

2010015873

ISBN 13: 978-0-7546-1184-4 (hbk)
ISBN 13: 978-1-138-25669-9 (pbk)

Contents

List of Figures

List of Abbreviations

ACAS	Advisory, Conciliation and Arbitration Service
AMB	Area Manpower Board
CDA	Co-operative Development Agency
CDP	Community Development Project
CTC	Charlton Training Consortium
CP	Community Programme
EDO	Employment Development Officer
ESF	European Social Fund
GERU	Greenwich Employment Resource Unit
GLC	Greater London Council
GLEB	Greater London Enterprise Board
GLMB	Greater London Manpower Board
GLTB	Greater London Training Board
ICTs	Information and Communication Technologies
ILEA	Inner London Education Authority
IMF	International Monetary Fund
ITBs	Industrial Training Boards
MSC	Manpower Services Commission
NHS	National Health Service
SOC	Standard Occupation Classification
TOPS	Training Opportunities Programme
YOP	Youth Opportunities Programme
YTS	Youth Training Scheme

Acknowledgements

First, I wish to acknowledge the debt I owe to former colleagues and friends at the Stonebridge Bus Garage Project and the Charlton Training Centre Project. My involvement with them, and the two community-based projects took place over a quarter of a century ago, but the memory abides. It is a happy memory of being part of exciting and ambitious attempts to address the social and economic problems of their communities. In telling my version of the stories of the two projects, I hope that I give justice to the events that unfolded, in a manner that accords with their own stories of those events.

I would like to thank Margaret Grieco, editor of the Voices in Development Management series, for her patient encouragement to bring this book to completion. My thanks also go to Carolyn Court and other staff at Ashgate Publishing for their help and support.

I am grateful to the London Transport Museum, NI Syndication, and Verso for granting permission to reproduce copyright illustrations and figures.

Finally, of all the friends to whom I owe a debt of gratitude, the greatest is Judi Holmes, my wife, whose patience and forbearance, support and love, have far exceeded any expectations to which I might have been entitled by our marriage vows nearly four decades ago.

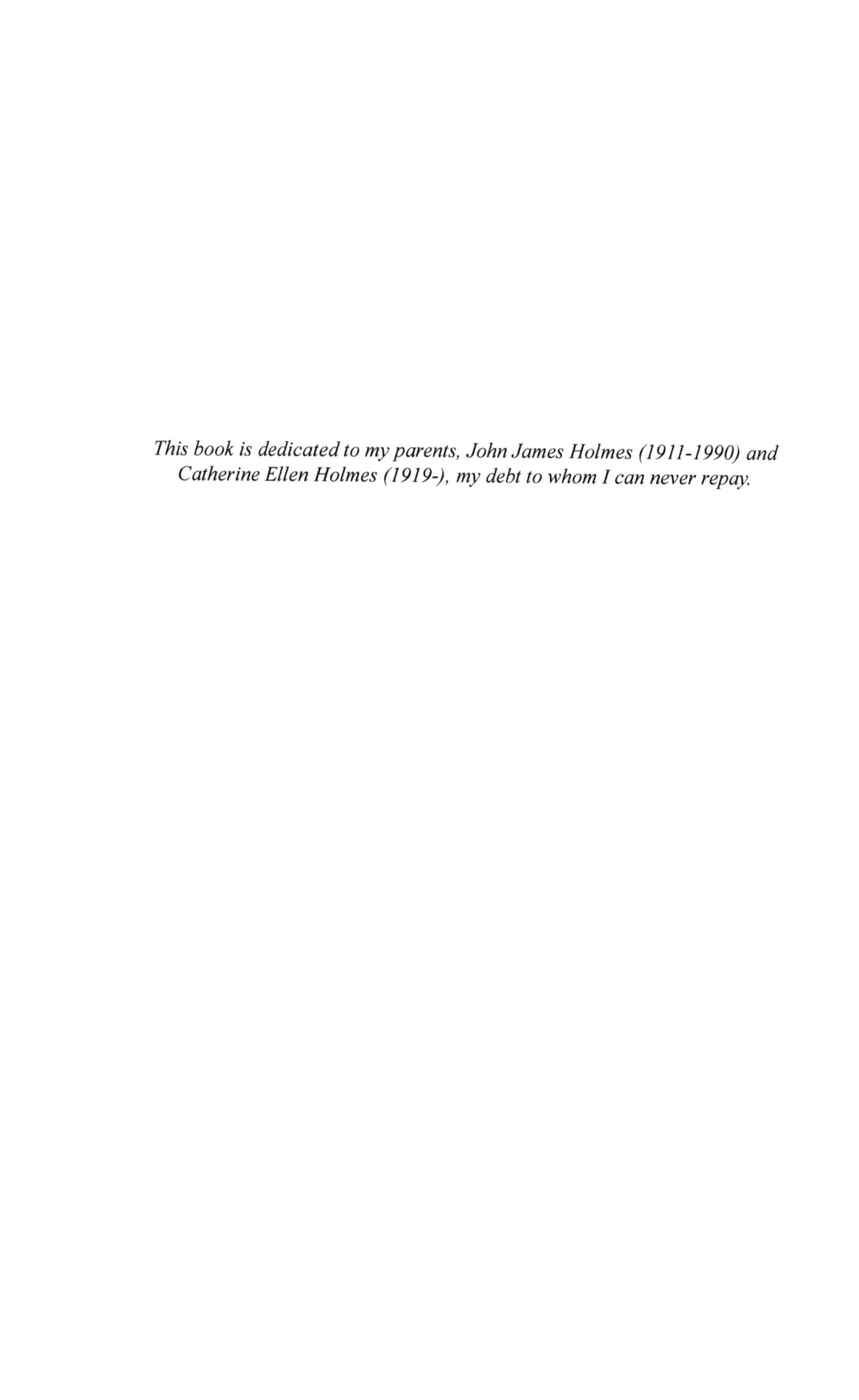

This book is dedicated to my parents, John James Holmes (1911-1990) and Catherine Ellen Holmes (1919-), my debt to whom I can never repay.

Chapter 1
Management: The Emergence of a Dominating Paradigm

The dominance of 'management' as a self-evident requirement of modern organization and institutional practice demands scrutiny. The decline of philosophies of participatory self-organizing in favour of 'managed' action, be it in the private, public or 'third' sectors, requires investigation. The economic, political, social, and technological circumstances which saw the emergence and development of both the practices and theory of 'modern management' have been radically transformed. In advanced industrial nations, where modern management emerged and developed, the economic and financial crises in the latter end of the first decade of the twenty-first century have severely challenged the promises of efficiency and effectiveness in promoting widespread economic and social wellbeing. Yet here and, in the context of increasing globalization, in communities in under-developed localities, the possibilities for alternative modes of organizing economic and social life are suppressed by the claims on technical and rational superiority of 'modern management'. This historically and geo-culturally located phenomenon, arising and developing within twentieth century industrial capitalist economies, continues to sustain its dominance, into the radically different context of the twenty-first century world.

At the end of the first decade of the twenty-first century, it is not difficult to find reasons to challenge the dominance of management. Corporate scandals such as WorldCom (Jeter 2003, Cooper 2008) and Enron (Elkind and McLean 2004, Eichenwald 2005) have received wide coverage, but are regarded as 'just the tip of the iceberg' (Elliott, Schroth and Elliot 2002). These, and the financial crisis that arose from 2007, resulting in unparalleled intervention by governments across the world, singly and jointly, are often attributed to various combinations of mendacity, greed, incompetence and poor regulation (Brummer 2009). However, questions may also be posed as to how companies that are deemed to be 'well-managed' at one time can, apparently, be transformed within a very short space of time into 'poorly-managed' organizations. Moreover, there have been financial consequences on other companies, in manufacturing, retail and other sectors, including the closure in the UK of iconic retailers such as Woolworths and Borders, that have little or nothing to do with the extent to which those firms were being managed in operational terms. Why were such possibilities apparently not foreseen and accounted for in conventional presentations of the theory and 'principles' of management?

Such conventional presentations place management as a phenomenon universal within organizations:

> We can say with absolute certainty that management is needed in all types and sizes of organizations, at all organizational levels and in all organizational work areas, and in all organizations no matter what country they're located in. This is known as the universality of management. (Robbins and Coulter 2005: 17-18, emphasis in original)[1]

Modern management arose within the growth of organizational society (Presthus 1979), its genesis (Pollard 1965) in the industrial revolution, developing as a definable set of practices as the scale of industrial enterprises expanded (Williamson 1975), progressively codified and articulated through an expanding body of what are now taken to be the early canonical texts by writers such as Taylor (1903), Gilbreth (1912) and Fayol (1949, but originally published in French in 1917). Early focus on the technical organization of work was supplemented by the end of the second decade of the twentieth century by attention to matters that were referred to as 'employment management' (Bloomfield 1920), focussing on 'human relations' at work, attention to such matters being developed further in the 1920s and 1930s (Roethlisberger and Dickson 1939, Urwick and Brech 1947). Subsequent developments of 'management thought' (Pollard 1974) or 'management theory' (Sheldrake 1996) are generally presented, often in textbook form, as a story of progressive discovery of key principles and development of successful practices. Within that story, management is viewed as essential to the effective and efficient operation of organizations, which are themselves regarded as bounded entities, rationally and technically superior to other forms of organizing economic and social affairs. Whilst initially a phenomenon of commercial activities, management and organization came to regarded as essential to other spheres of collective human endeavour, including healthcare, education, public administration. Organizing, organization, managing and management (and managers) became inseparable.

Alternative Practices

The conventional, dominant story has not gone unchallenged. The place of management in the capitalist economic and social system has been a major source of challenge from those who seek the overthrow of the system, rather than merely its amelioration. The collapse of communism in the former USSR and its satellites, at the end of the twentieth century has somewhat muted that source of challenge,

1 This text by Robbins and Coulter was published in its eighth edition in 2005, having been first published in 1983 with Robbins as sole author. The Preface to the eighth edition claims that the book is 'the number-one selling basic management textbook in the United States and the world!' (exclamation mark in original).

as a viable alternative system can only be a vague prospect rather than shown to be presently enacted. Challenge has increasingly developed under the broad banner of antiglobalization (Klein 2000, Derber 2002, Cohen and McBride 2003, Brysk and Shafir 2004, Polet 2004).

Yet even during the ascendancy of conventional management (that is, traditional capitalist) practice during the twentieth century, attempts at organizing collective economic and social activities in an 'alternative' manner have occurred at various times and in various locations (Lindenfield and Rothschild-Whitt 1982, Castells 1983, Young and Rigge 1983, Jenkins and Poole 1990, Parker, Fournier and Reedy 2007). As Parker, Fournier and Reedy show, 'there are many alternatives to the way that many of us currently organize ourselves' (that is, market managerialism) (2007: ix). In the context of economic activity, such attempts would generally be concerned with issues of ownership and control, and be presented as 'alternative' by the use of such terms as 'co-operative', 'industrial democracy', 'workplace democracy', 'workers' control' (Coates and Topham 1968). Various forms of co-operatives exist, demonstrating the ability of groups and communities, of varying sizes, to engage in their common affairs (commercial, such as producer and consumer co-operatives; social, such as community facilities, housing), through democratic self-organizing and common or shared ownership of resources (see Co-operatives UK for UK examples). Although often relatively small, the Mondragon Corporation in Spain has provided an example of such self-organizing on a large scale for over half a century (Bradley and Gelb 1983).

In the wider social sphere, particularly within urban areas, there have been many experiments in community-based participatory approaches to the planning and organizing of neighbourhoods, for provision of housing, social, educational, healthcare, recreational facilities and services (see Castells 1983). During the late 1970s and 1980s, a number of such experiments were developed under the notion of 'popular planning' (Benington 1986, Geddes 1988, Brindley, Rydin and Stoker 1996).

However, frequently these community attempts to provide an alternative to traditional management structures encountered problems; experiments were subverted by a number of interests and dynamics. This book provides an account of two radical experiments within the popular planning movement of the United Kingdom. These commenced from icons and images of self-organizing by disadvantaged groups and communities but were subverted by the imposition, importation and adoption of traditional management approaches. In providing the account here within a framework for critiquing dominant representations of modern management, the dominance of management, the aim is to avoid merely 'blaming the victims' (Ryan 1976). Rather, the aim of the 'autopsy' of these experiments is to provide an insight into the dynamics likely to affect community and popular planning projects within the context of development management in the developing world. In providing an analysis of the processes of subversion experienced by these initially self-organizing experiments within a culture and a structure of the dominance of management, it is hoped that this book provides a set of tools for exploring the subversion of participatory projects within the development context of the dominance of international development aid organizations.

Alternative Theory

Whilst the literature on and about management has been dominated by the conventional model, as discussed, this has been accompanied by literature that presents analyses critical of that conventional model. From the late 1960s to the early 1980s, there was a growing body of 'radical' and 'critical' sociological examinations of organization and management (for example Nichols 1969, Beynon 1973, Braverman 1974, Clegg and Dunkerley 1978, Burrell and Morgan 1979, Clegg and Dunkerley 1980, Clawson 1980, Zey-Ferrell and Aitken 1981). Braverman's (1974) re-interpretation of 'Scientific Management' as the outworking of Marx's analysis of the capitalist labour process led to what became termed 'Labour Process Theory', the source of considerable academic endeavour, particularly through the 1980s and 1990s (Friedman 1977, Littler 1982, Storey 1982, Wood 1982, Knights and Willmott 1989). More recently, analyses have been undertaken within what is now broadly termed 'critical management studies' (Alvesson and Willmott 1992, Grey and Willmott 2005, Alvesson, Bridgman and Willmott 2009b), which has 'emerged as a movement that questions the authority and relevance of mainstream thinking and practice' (Alvesson, Bridgman and Willmott 2009a: 1).

Critical Management Studies is certainly not a unitary field; it is, rather, 'a pluralistic, multidisciplinary movement incorporating a range of perspectives' (Alvesson, Bridgman and Willmott 2009a: 5). Within such a 'broad church' (Grey and Willmott 2005: 14), a number of common concerns and themes are shared. These include the questioning of the assumption that modern management is a positive and socially valuable approach for organizing economic and social affairs, that it has developed as a social practice because it is rationally and technically superior to other (particularly collective) modes of organizing. Such questioning calls attention to the power relations that are in play, modes of domination that arise and are enforced, inequities that are generated, through the functioning of the supposedly-disinterested operation and application of management principles. An overarching theme is that management has arisen historically within, and reflects and promotes, the development of capitalist society.

This book aims to be a contribution to such 'radical' and 'critical' studies of organization and management. Any contribution will be just one voice amongst a wide range of others. However, to date the main type of contribution has been primarily theoretical in nature, particularly those that draw upon and seek to promote a particular theoretical perspective and those that address particular topics (such as globalization, power, identity, gender, culture) and/or particular specialism within management (marketing, accounting, human resource management, and so on). There is very limited presentation of empirical research undertaken from a radical/critical perspective. Moreover, where a radical/critical perspective is adopted for the analysis of management within particular settings, these tend to be 'mainstream' organizations in the private or public sectors. Less attention has been paid to situations in which deliberate attempts have been made to engage in 'alternative' modes of collective organizing of economic and social affairs.

This book therefore seeks to contribute to such studies by presenting and examining two contexts of such 'alternative' approaches. These were organizations espousedly intended to be in some sense 'alternative' organizations, being established by community organizations on non-profit distributing principles. One organization was originally intended to be developed as a 'community co-operative'; the other was intended to 'promote democratic structures of control'. As will be seen both organizations encountered severe problems, one of them eventually ceasing. In attempting to understand the nature of management in these organizations, it is necessary, I shall argue, to examine the societal processes which are generated within the capitalist mode of production and which penetrate and permeate all forms of organizations.

A Tale of Two Organizations

Stories Worth Telling

Brief descriptions are provided here of the two organizations in which the events described and analysed took place. The purpose of this is to enable the reader to gain an overview of the complex nature of the unfolding history of each organization so that the more detailed descriptions in the later chapters might therefore be easier to grasp. The period in which the events depicted took place was, in both cases, about four years in the early- to mid-1980s.

In presenting the accounts of the two organizations, the actual names, rather than pseudonyms, are used. The use of pseudonyms would be rather fruitless, given the high profile that each had and the fact that there were unique characteristics that would make each of them obvious from the descriptions. Moreover, in telling the story of the two community-based initiatives, my hope is that the attempts by the key persons and communities may be a source of celebration. The situations faced by those who initiated and participated were acute and pressing, and conventional efforts to deal with them were largely ineffectual. Taking steps to start to address their situations, to continue with their efforts during periods of considerable difficulties, took considerable amounts of courage and hope, and *should* be celebrated.

However, as is conventional in research publication, individual persons involved will *not* be named, but generally referred to by the positions they held, where the position titles are capitalised (for example, 'Project Co-ordinator'). Although the previous paragraph referred to 'the accounts', it should be noted that this book contains just one version of such accounts, one story, presented by myself, the author. Each person involved in and with the two organizations will have their own story, their own account of the events that unfolded, each of which may agree or divert from my own in various ways. My hope is that in telling *my* story, I have in some way been able to do justice to the visions and hopes that gave rise to the efforts by those who participated, and presented those efforts in positive terms, so as to avoid simplistic explanation of the difficulties experienced in terms of what Ryan (1976) calls, evocatively, 'blaming the victims'.

As stated, the events described and discussed here occurred some two and half decades ago, something that requires comment. First, the passage of time reduces any possible effects of adverse perceptions of those involved. The basis on which the descriptions and analyses of the events are presented here is not that any blame should be apportioned to any individuals or groups. Indeed, the argument here is that to understand the histories of the two organizations, we need to move to a level beyond that of the actions of individuals, to consider how structural features of our contemporary society play a determining part, specifically in respect of the dominance of management.

A second comment that should be made is that the story, or stories, of these two organizations has had little opportunity for expression and communication. The voices of those involved have, for the most part, been muted. This volume is one within a book series that seeks to provide an opportunity for the voices of participants in 'grass roots' organizations to be expressed and amplified. The importance of 'storytelling' in the field of development management is now being recognized more widely (Denning 2000), just as it is increasingly so in the field of organization studies (Czarniawska-Joerges 1998, Gabriel 2000, Boje 2001). By telling the story, or at least, *my* story of these two organizations, other groups may find resonance with their own situations and find pathways towards maintaining their founding visions within the pressures to conform to the dominance of management.

The third comment follows from the previous. Given the economic and financial crisis affecting much of the 'west', the developed world, it is almost certain that governments will seek to address the continuing economic and social problems facing their populaces by looking to solutions based on 'local community' initiatives. Yet often such local community initiatives are developed with radical intent, in terms both of goals and of organizational form. The lessons from history should therefore not go unheeded, if such initiatives are to be truly community-based, rather than be transformed and incorporated into 'mainstream' organizational forms as has often been the case (Zald and Denton 1963).

As indicated above, the story presented here of each of these two organizations is *my* story. In the case of the first, the Bus Garage Project, I was employed as one of the 'professional' staff of the Project and so was party to many of the events described. This included access to a wide range of documentation, including minutes of meetings, reports, and correspondence. Whilst parts of the story are based on my observations, most is supported by or reflected in such documentation.

Explicit approval was sought from and granted by the Steering Group for research to be undertaken.

In the case of the second organization, the Charlton Training Centre, I was employed as an external researcher by a polytechnic commissioned to undertake a research programme, which covered other areas of study undertaken by other research staff. The research project took place over a 15 month period from early 1985 to spring 1986, much of which time was spent in participant observation at the Centre. As with the Bus Garage Project, a considerable amount of documentation was also examined. This enables 'thick description' (Geertz 1973), providing for detailed accounts of the two histories *and* presentation of key aspects of the context understanding of which, it will be argued, is necessary for full understanding of the stories of the two organizations.

The Bus Garage Project

The first organization examined is the HPCC Bus Garage Project. The Project was established in 1981, at the site of a disused London Transport bus depot on the edge of the high-rise Stonebridge estate in the London Borough of Brent. The purpose of the Project was to convert the bus depot into a large community centre to provide for the local community social facilities (such as leisure, sports, welfare, educational and training) and economic facilities (enterprises which create jobs and bring about local economic development). At the time, the neighbourhood was a deprived area with a high ethnic minority population; over 70 per cent of the local people were Afro-Caribbean Black and some 10 per cent were Asian.

The problems typically associated with deprived inner city areas were particularly acute within this area at the time of the events examined in this book: very high rates of unemployment, especially among young black people, lack of community facilities, high rates of crime, high rates of custodial sentences passed on young males convicted of crimes, poor educational achievement, and so on. The origins of the Project lie in the response of some of the local young black people to the pressures they were experiencing in early 1981. Unlike other areas with similar problems no riots took place. Instead a group who were prominent in the area, who may be seen as typical of the local young black people, formed themselves into a community action group and began to press for more resources and facilities for the area. With help from the local council a variety of activities were initiated.

Later that year London Transport closed the bus depot on the edge of the estate. The depot was near to office blocks on either side of the busy North Circular Road, one of which adjoined the local British Rail and London Underground station. The local group, now calling itself a 'Community Council', successfully persuaded the local authority, the London Borough of Brent, to undertake a feasibility study of converting the bus depot site into a facility for the local community, rather than allowing it to be redeveloped into yet another office block. A steering group was established to undertake the feasibility study, with six local councillors, four members from the Community Council, and two representatives from other local community, forming the steering group. They were assisted by officers of the local authority, and workers at the local Cooperative Development Agency (CDA) and Community Law Centre.

The steering group completed its report in December 1981, with a foreword written by the Chairman of the Community Council who was also chairman of the steering group, and a preface written by the Leader of the Council. The report stated that

> It should be made clear at the outset that the Bus Depot Project is not proposing that the Council, Central Government and other agencies convert the Depot into a multi-million pound community Project and sports complex. The philosophy behind the Project is totally different. The starting point for the project is the desire of the people of Stonebridge, as expressed by the Harlesden People's Community Council and other local community groups, to improve local conditions *through their own efforts and in their own way*. (Stonebridge Bus Depot Steering Group 1981: 4, emphasis added)

The report proposed that the site be bought by Brent Council and converted to house a variety of activities and facilities, including sports and leisure, education and training, workshops for local enterprises, offices for running the Project.

The cost for acquisition of the site was estimated at £2 million; the costs for conversion were estimated at over three quarters of a million pounds. Other areas of cost included capital expenditure for individual activity areas of an estimated £350,000 and support for the running expenditure of over half a million pounds per year initially, declining over time. Various sources of funding proposed included the Departments of Environment and of Industry, the Home Office, the Manpower Services Commission, the Greater London Council, various charitable trusts, and the European Social Fund. Brent Council was considered to be able to provide capital funding for purchasing and converting the site, but 'revenue' funding was considered to be improbable because of budgetary constraints. Finally, it was proposed that the Project as a whole should be run by a 'community cooperative' open to anyone who lived in the area, or who were members of individual cooperatives in the Project, and who agreed with the declared aims of the community cooperative. The community cooperative would elect a management committee to direct and control the Project.

The proposals were accepted by Brent Council in March 1982, with support from all three political party groups represented on the Council. With funding support from the Greater London Council, Brent Council purchased the bus depot, giving a specially formed Steering Group (replacing the original steering group) a licence for the site. Successful applications were made to various local and national government agencies and departments, and also the European Social Fund which provides funding for vocational training initiatives. The Project was visited by various well known political figures, who espoused enthusiastic support for the project.

The first phase of redevelopment was completed in late 1983, and officially opened amid widespread acclaim. However the first phase was only a small part of the overall plan, occupying a separate building from the main area, the 'main shed', and the redevelopment had been largely managed by the local authority. By the end of 1983 the estimated costs of the plans drawn up by the Steering Group had risen to over £5 million. The start of work on the main shed, planned to start in early 1984, was delayed because of a shortage of funds and the various local and national authorities involved began to question the plans. By the middle of 1984 the Project was in deep crisis in terms of its internal affairs, in respect of its support from the authorities, and in terms of its support within the local community.

A consultant was brought in by the Department of the Environment. His report was very unfavourable, but by the time he presented his report the link officer from Brent Council (the Policy Coordinator, who was later redesignated 'Assistant to the Chief Executive') had persuaded the Steering Group to re-examine its plans. After further discussions and negotiations the Steering Group were persuaded to dismiss the firm of architects and to engage on a contract basis the Development Department of Brent Council. Several months later a new scheme design and set of plans were approved. The Department of the Environment approved the funding but insisted that its dealings should be only with Brent Council, not the Project directly.

The Project was completed in 1988, now named 'Bridge Park'. It was officially opened by His Royal Highness The Prince of Wales, to great celebrations and media coverage. However, over the next few years, the income to Bridge Park did not meet the levels anticipated, and Brent Council continued to have concerns of the Centre's management. Bridge Park itself was still leased only under licence from Brent Council, which could be ended with little notice. When Brent Council attempted to end the licence in order to regain control of the Centre, a series of court actions ensued, ultimately concluding in 1992 in success for Brent Council, and the end of the Community Council's involvement. Bridge Park still operates today, primarily as a leisure centre run by Brent Council, with rented premises for local businesses.

These events will be examined in two parts; these will cover a time period of around five years, from 1981 to 1986. Chapter 8 will examine a dispute between management and workers employed at the Project funded by means of the MSC's Community Programme, a short-term employment programme for long-term unemployed people. The processes of domination by management of the very local people for whom the Project was ostensibly initiated will be analysed and related to wider societal processes. Chapter 9 will examine more broadly the history of the Project in this period, and will question the appearance of the Project as a community initiative. On the contrary it will be argued that the events, and the nature of management of the Project, must be understood in terms of the activities of the state in support of capital at a time of crisis.

Figure 1.1 Bus Depot in 1981 shortly before closure
Source: © TfL from the London Transport Museum, photo by Peter Wilson.

Figure 1.2 Members of Community Council in 1983 after project started
Source: © *The Times* 3 March, 1983/nisyndication.com.

Figure 1.3 HRH Prince of Wales opening Bridge Park in 1988

The Charlton Training Centre

The second organization studied (Chapter 10 and 11) is the Charlton Training Centre. This was based at the premises of a former Manpower Services Commission (MSC) SkillCentre in south east London, which was closed by the MSC in 1983. The decision by the MSC to close the SkillCentre was made after a short 'consultation' period, during which a campaign was established to save the SkillCentre. That campaign, launched by the local Greenwich Employment Resource Unit and supported by both the Greenwich Council and the Greater London Council (GLC), as well as other local community groups, then turned into a campaign to establish some form of replacement for the skills training facilities which were to be lost. This was the origin of the Charlton Training Consortium. The development of proposals soon became concentrated on taking over the SkillCentre site and operating it as a skills training centre under some form of community direction and control. A development worker was employed with funding from the GLC and detailed proposals began developing. These proposals included the establishment of a unit for women only, both trainers and trainees, within the centre.

Funding for this unit was sought and obtained from the European Social Fund, matching funds committed by the GLC. In December 1983, the GLC agreed the funding application. By April 1984 the Consortium had taken over the lease of the SkillCentre site and begun to recruit staff. Details of proposed courses were developed, and arrangements for other aspects of the Centre's operations established. The Centre was to engage in a positive action programme, seeking to provide access for people who are disadvantaged in access to training: black and other ethnic minority people, women, disabled people, gay men and lesbian women. Outreach was to be undertaken to encourage applications for training places from such groups; childcare was to be provided to enable those with such responsibilities to take up training places; educational provision was to be made to assist those who wished to improve their literacy and numeracy.

Moreover the Centre was to be managed in a 'non-hierarchical' manner, involving workers, trainees and the local community in the direction and control of the Centre's operations. The first courses started in December 1984. In March 1985 the GLC approved funding through to December 1985. More courses were established and trainees recruited. Later in 1985 the future funding of the Centre was examined, and early in 1986 the GLC agreed to fund the Centre until the end of September 1986. The GLC was itself abolished in March 1986, and responsibility and powers for cross-borough grant-aid for the voluntary sector passed to the newly formed London Boroughs Grant Committee. In September 1986 an application from the Consortium to that Committee was not approved. The Committee decided that further negotiations should take place 'for the re-establishment of a training facility in South East London at a realistic level of funding and with less dependence upon public subsidy'.

Discussions leading up to this decision had taken place between the officers of the London Boroughs Grant Scheme and officers of the Greenwich Council. The conclusions they came to were put to the Consortium. They included establishing the post of overall manager in place of the collective responsibility of four coordinators, each responsible for a particular aspect of the Centre's activities and with a team of staff reporting to them. The officers also proposed a restructuring of the Consortium to provide a controlling position to the funding boroughs, with representatives of the community acting in an advisory capacity. At the end of September the Centre closed and all staff were made redundant.

The actual management structure of the Centre was a continuing source of problems. The Memorandum of Association of the Charlton Training Consortium stated that '[i]n carrying out the aforesaid objects the association ... will promote democratic structures of control in the work place and participation in those structures by all employees and trainees'. The structure never really operated as it had been agreed, and on several occasions the Board discussed the organizational structure and management of the Centre, without reaching clear consensus. The main funder, the Greater London Council, was abolished in March 1986, but continued funding was being considered by Greenwich Council and the London Boroughs Grants Scheme, the body established to provide finding to voluntary sector organizations in London. However, the management structure was regarded by officers of both the Greenwich Council and the London Boroughs Grants Scheme as being a major problem, stating that '[i]t is vital if Charlton is to continue to function, that the management of the centre is put on a firmer footing'. The proposed remedy was to introduce a hierarchical structure under the direction of a board which was controlled by local authority nominees. This proposal was still under active consideration by those officers after the closure, to enable the Centre to be re-opened by the local authorities concerned; however, the Centre never re-opened.

These events will be examined in two stages. Chapter 10 will examine the influences of the state, through the two local authorities involved and the Manpower Services Commission, which significantly determined the development of the Centre. The effects of these influences were of such importance that the notion of the Centre being under democratic control by the trainees, the staff and the community may be seen to be problematic. Chapter 11 will examine the attempts by those most closely involved within the Consortium and Centre to develop an organizational form which met the intentions stated in the Memorandum of Association.

Chapter 2
The Nature of Management

What is Management?

The question 'what is management?' seems a simple one, yet is deceptive. There is a tension in much of the debate about management concerning what it is. For some, management is 'what managers do during their working hours' (Lupton 1983: 17). For others, management, or managing, is presented as a common and universal human activity, but one that has become increasingly important in advanced societies such that it has become the particular responsibility of a distinct occupational grouping members of which may and should develop their knowledge and skills as managers. Thus, in the tenth edition of a popular international textbook on management, Weihrich and Koontz[1] state that managing is

> [o]ne of the most important human activities. ... Ever since people began forming groups to accomplish aims they could not achieve as individuals, managing has been essential to ensure the coordination of individual efforts. As society has come to rely increasingly on group effort, and as many organized groups have become large, the task of managers has been rising in importance. (Weihrich and Koontz 1993: 5)

In similar vein, Boddy writes that, '[a]s individuals we run our own lives, and in this respect we are managing. Management is both a general human activity and a distinct occupation' (Boddy 2002: 11, emphasis in original). Some restrict their claim that on the universality of management to organizational contexts; for example:

> We can say with absolute certainty that management is needed in all types and sizes of organizations, at all organizational levels and in all organizational work areas, and in all organizations no matter what country they're located in. This is known as the *universality of management*. (Robbins and Coulter 2005: 17-18, emphasis in original)[2]

---

1 This textbook started life in 1955 as *Principles of Management: An Analysis of Managerial Functions*, co-authored by Harold Koontz and Cyril O'Donnell. Its longevity testifies to its key place in the development of management as a subject for study in further and higher education in the latter part of the twentieth century.

2 This text by Robbins and Coulter was published in its eighth edition in 2005, having been first published in 1983 with Robbins as sole author. The Preface to the eighth edition

Such a claim to 'absolute certainty' warrants critical scrutiny.

Of course, it might be argued that we might expect to find such claims in textbooks written for students on vocationally-oriented programmes; after all, they are published within the increasingly commercialized and highly competitive educational marketplace. Yet similar views have been expressed by authors writing mainly for an academic, rather than student, audience. In what is presented as a critical perspective on management, Alvesson and Willmott state that

> In all societies, people are involved in the complex and demanding business of organizing their everyday lives. Each of us engages in a daily struggle to accomplish ordinary tasks and maintain normal duties. This management of routines is something that we all contribute to, and are knowledgeable about – it is 'second nature'. (Alvesson and Willmott 1996: 10)

Admittedly, the key point the authors seek to make is that 'in modern societies', much of everyday life has become subject to shaping, organizing and regulation by experts, including managers. However, here is a danger of conceptual confusion arising from this way of talking about management in 'everyday lives' in the same context as engaging in the analysis of the social phenomena referred to by concepts of management and organizations that characterize much of modern society. It is therefore vital to engage in some analysis of the concepts of management and managing, for, as Andreski warned, 'constant attention to the meaning of terms is indispensable in the study of human affairs, because in this field powerful social forces operate which continuously create verbal confusion ...' (Andreski 1972: 61).

Definitions and Differences

Textbooks on any subject often seek to provide a definition of a key term, which in our case is 'management'. In doing so, they usually neglect to consider the way that the term is used in actual practice and to notice how there may be differences in use, and so of meaning. The danger then is that this may impair our efforts to understand the phenomena about which, and the contexts within which, the term is used. Our intellects may be 'bewitched by means of language', as Wittgenstein (1953: para. 109) put it. Recognition that a term may have different meanings in different contexts of use goes back at least as far as Aristotle. Flew uses the term 'systematic ambiguity' to refer to the situation that arises when 'words or expressions that may always have the same meaning when applied to one kind of thing, but have a different meaning when applied to another kind of thing' (Flew 1979: 11). Such 'different meaning' does not necessarily mean *unrelated*

---

claims that the book is 'the number-one selling basic management textbook in the United States and the world!'

meaning: quite the opposite. Indeed it is often the closeness in meaning that causes the confusion, or 'intellectual bewitchment'. So we shall first consider the term 'management' in respect of such possible systematic ambiguity.

Misplaced Categorization: Mundane versus Abstracted Management

The claim that managing is a common and universal aspect of human life, that 'we all manage our everyday activities' is a particular source of conceptual confusion. We do indeed use words based on the stem manag(e) – in everyday, mundane discourse about aspects of everyday life; for example:

- 'I just managed to get to the airport on time.'
- 'Can you manage those stairs on your own?'
- 'She managed to raise four children by herself and still pursue her career'.
- 'We can't expect a retired couple to manage on the basic state pension'.

However, it would be invalid to conclude from such use of the term that people ordinarily, in everyday life, engage in the same kinds of activities as are undertaken by those who are formally designated as 'managers' in organizational settings. This can be demonstrated by asking what is done by the use of the word 'manage', in addition to what is done without that word (that is, 'I got to the airport on time'; 'Can you climb those stairs on your own?'; 'She raised four children etc.'; 'We can't expect a retired couple to live on the basic pension'). The additional work done by using the word 'manage' would seem to draw attention to the fact that some (ordinary) activity is accomplished despite some difficulty or other; or (in the final example), we should expect certain difficulties (insufficient income) to prevent a certain accomplishment (living the kind of life to which a retired couple should be entitled). This 'drawing attention to' may follow on from certain information previously given or jointly known; or it may implicitly evoke a question concerning the nature of the difficulty; or perhaps it draws upon common understanding about what may be realistically expected. Used in such mundane modes of discourse, the term 'manage' also generally serves to provide some emotional tone, for example, of relief, (self-)congratulation, concern, admiration, justifiable anger, or whatever.

It would therefore be invalid to argue that, because people ordinarily manage various activities in their daily lives, they are therefore 'managing', engaging in 'management' *in the same sense* as those terms when they are applied to managers in organizations. To do so would be based on a *misplaced categorization* of those mundane activities.

Management as a Social Accomplishment

The mundane use of the term 'manage' in the sense of individuals accomplishing some desired or desirable outcome in the face of difficulties should also be distinguished from its use in the context of the co-ordination and control of collective activity. It is the latter that applies in the case of management in and of formal organizations. Here again we may face conceptual confusion because, although there may be individuals, 'managers', who are said to be 'managing', the social accomplishment of the management, that is, co-ordination and control of activities towards certain ends, requires that those who are being managed must play their part. Management may thus be said to be a joint accomplishment, by managers and managed, in line with what Giddens (1984) terms the 'dialectic of control'.

However, it does not follow from this that we can say that those who are managed are engaging in managing in the same sense as we might say a manager is engaging in managing. There are two problems here. First, such an inference would be to make a category mistake (Ryle 1949), that is, to confuse different logical types of concepts. It would be akin to concluding that, because a football team had won a match with a score of two goals to nil, each player had individually won the match, and each had scored two goals. The logical grammar (Wittgenstein 1953) of the concept of management of a department, of a project, of an organization, and so on, as a social accomplishment differs from that of the activities in relation to certain categories of individuals who spend their daily working lives in the dominant form of contemporary work organizations.

A second problem is that 'to manage' may be used as both a *task* verb and an *achievement* or *success* verb (Ryle 1949). Ryle uses this distinction to argue that certain terms in the latter category, because they are active verbs, tend to make us oblivious to their logic, citing examples such as 'win', 'unearth', 'find', 'cure', 'convince', 'prove'. These correspond with the related task verb with the force of 'trying to'; sometimes we use an achievement term as a synonym for a task term, for example, 'mend' as a synonym for 'try to mend'. The main difference between the logical force of a task verb and its corresponding achievement verb is that, in using the latter 'we are asserting that some state of affairs obtains above that which consists in the performance, if any, of the subservient task activity' (Ryle 1949: 143). In the case of 'manage' and 'management' as a social accomplishment, that is, the outcome of the collective combination of activities on the part of both manager(s) and the managed, we can see that the term is being used as an achievement or success term. The success arises from the various actors undertaking *separate* activities in what thereby becomes the collective accomplishment.

Descriptive and Evaluative Conceptualizations of Management

A further consideration of the different meanings that the terms 'management' and 'managing' may have is that sometimes the terms have some form of evaluative connotation, whereas at other times they are used in a more neutral descriptive sense. The evaluative sense is often found in the mundane use of the terms considered above, that is, where such use connotes some source for congratulation or celebration ('you managed it – well done!'). Consider, however, the following discussion by Hales:

> In general, when human beings 'manage' their work, they take responsibility for its purpose, progress and outcome by exercising the quintessentially human capacity to stand back from experience and to regard it: prospectively, in terms of what will happen; reflectively, in terms of what is happening; and retrospectively, in terms of what happened. Thus management in its general sense is an expression of human agency, the capacity to shape and direct the world actively rather than simply react to it. (Hales 2001: 2)

The passage starts as what appears to be a description: this is what managing is; however, the general tone of the text is clearly evaluative. Thus humans 'take responsibility', exercising their 'quintessentially human capacity'; the opposite of managing is 'simply' reacting to the world. Rather than seeing this as a form of description of managing, Hales is using the term 'manage' as an evaluative concept which is used in and of those kinds of situations in which the way those persons who are involved are conducting themselves is generally held in high regard.

Hales continues, immediately after the above passage, arguing that the process of management

> therefore subsumes five conceptually distinct, if, in practice, interrelated elements. To manage work in general means:
> *Deciding/planning* what is to be done and how.
> *Allocating* time and effort to what is to be done.
> *Motivating*, or generating the effort to do it.
> *Co-ordinating* and combining disparate efforts.
> *Controlling* what is done to ensure that it conforms with what was intended. (Hales 2001: 2)

He appears to present this as some form of analysis of what is now termed 'the process of management', into a set of 'elements' that are 'conceptually distinct'.

However, two problems arise with such an approach. First, Hales provides no argument to warrant his use of 'therefore', purportedly linking the two passages. Secondly, and more importantly, Hales introduces a set of categories (deciding/ planning, allocating, and so on) that are themselves problematic. Whilst initially appearing to be, and deployed by Hales as, *descriptions* of activities (or 'task verbs' in Ryle's terms) that together constitute managing, they might better be viewed as semantic elaboration of what is itself an *evaluative ascription* (or 'success verb'). To say that someone decided and planned what should be done, or generated the effort to do it, provides no further information describing what activity that person engaged in, merely that, as a result of some (unspecified) activity, 'what should be done' was decided and (at least the start of) some effort was put into doing it. We have no information about the manner in which the decision was manifested: for example, just 'in the mind' of the person, committed in writing to the back of an envelope, stated orally to someone else, published on the internet, or any of a host of other possible ways. We are provided with no criteria for distinguishing between deciding/planning and not-deciding/planning *as activities*, that is, no indicators that would tell us that someone is deciding/planning and another is not.

Is Management What Managers Do?

The foregoing discussion emphasized the importance of distinguishing between descriptive and evaluative conceptualizations of management. Lupton's (1983: 17) notion that management is 'what managers do during their working hours', if valid, could only apply to descriptive conceptualizations of management, where 'management' is effectively synonymous with 'managing', and where 'managing' refers to an activity, or set of activities carried out by managers. However, two problems arise with this formulation. First, it raises questions about how we are to use the term 'manager', that is, who may be so categorized and on what basis? Secondly, it assumes that there is an exclusive, or mostly exclusive, set of activities common to the work of such managers.

Management as an Occupation

One approach to the first issue is to regard management as an occupation, where a person's job title would clearly place them into the category of manager. The job title might not include the word 'manager', but may have a term, such as 'executive' or 'director' that might be taken as indicating a management position. However, job titles vary considerably; the actual wording may not be clearly indicative of what is involved so there are bound to be boundary difficulties in terms of whether a specific job is to count as a management position. Nevertheless, attempts have been made, for particular research purposes, to determine what does and what does not constitute a management job; the most thoroughgoing approach is that of government statistical agencies, where continuity and consistency of definition over extended time periods are deemed to be essential.

In the UK, the Office for National Statistics conducts the Labour Force Survey, in each of four quarters of the year, providing within that its estimates of the numbers in various occupational categories. The occupational categories are defined by the Standard Occupational Classification (SOC) system[3] (itself subject to revision from time to time, the last occasion being 2000), which places

3 The United States of America also has its own Standard Occupational Classification system, as do a number of other countries; the International Labour Organization developed the International Standard Classification of Occupations.

all occupations into nine major groups. Group 1 has the title 'Managers and Senior Officials', which

> covers occupations whose main tasks consist of the direction and coordination of the functioning of organizations and businesses, including internal departments and sections, often with the help of subordinate managers and supervisors. Working proprietors in small businesses are included, although allocated to separate minor groups within the major group. (Office for National Statistics 2000: 37)

On the basis of the SOC definition, the Labour Force Survey for the last quarter of 2006 put the estimate of the number of managers in the UK as just under 4.4 million (Office for National Statistics 2006: S28). Other estimates of the number of managers in Britain vary dependent on the definitions used, ranging from 2.5 million to over six million (Council for Excellence in Management and Leadership 2002, see also Williams 2001, Burgoyne, Hirsh and Williams 2004). On any estimate, the population of the UK managerial workforce is sizeable, the Labour Force Survey showing it to have grown from 13.6 per cent of the total workforce in 2001 to over 15.5 per cent in 2009.

This growth in the size of the managerial labour force over the first decade of the twenty-first century continues the trend of the twentieth century. Price and Bain (1988) indicate a four-fold increase in the percentage of the working population engaged in work as managers over the seven decades between 1911 and 1981, with the percentage more than doubling over the previous two decades. That rapid growth contrasts with the relatively slow growth during the previous century as the industrial revolution proceeded apace (Pollard 1965, Mant 1979). Management as an occupation is largely a twentieth century phenomenon.

Yet management is by no means a unitary occupation. Within the UK's Standard Occupational Classification (SOC) system (Office for National Statistics 2000), the 'major group' of 'managers and senior officials' is sub-divided into 13 'sub-major groups' including 'corporate managers and senior officials', 'production managers', 'quality and customer care managers', 'managers in hospitality and leisure'. Clearly the range is very wide. Moreover, it does *not* include head teachers, who would be classified under a different 'major group', that of 'professional occupations', despite the increasing emphasis upon the role of head teacher as the *manager* of a school. Of course, any classification system for application to the social world is likely to yield problems of this kind, and specific systems will generally resolve them in terms relevant to the purposes for which they have been established. In our case, this points to the need for caution in adopting the Lupton formula that 'management is what managers do'.

Heterogeneity of Management

We certainly cannot assume that managers (as categorized) in the early years of the twenty-first century, or even in the latter years of the last century, engage in the same activities as those at the start of the last century. Indeed, we can safely assume that the activities of managers over the past century have changed considerably. The whole thrust of the early 'scientific' management writers was to *promote* change in the activities that managers undertook, and the subsequent development of the body of management literature (see for example Pollard 1974, Crainer 1999) has continued in that vein.

The initial codification of the activities of management is generally credited to Fayol who stated that to manage is 'to forecast and plan, to organize, to command, to co-ordinate and control' (Fayol 1949: 6). Such an approach was disseminated in the English-speaking world mainly by Urwick, through his book *The Elements of Administration* (Urwick 1943) and his consultancy work. Variations, elaborations and modifications of what is generally referred to as the 'classical school of thought' or 'classical view' in management, may be found in standard management textbooks that have appeared since the Second World War. However, such attempts at codification were, in themselves, not based on empirical observations of what managers actually do.

Where there have been observational studies of the actual behaviour of managers (Stewart 1967, 1976, Mintzberg 1973, Kotter 1982), the classical model of management, presented as a set of rational activities, has been seriously challenged. Stewart's studies identified considerable *variety* between managers in the nature of their work activities, which was characterized by brevity and fragmentation. Mintzberg is scathing of the 'classical' view of managerial work:

> If you ask a manager what he does, he will most likely tell you that he plans, organises, co-ordinates, controls. Then watch what he does. Don't be surprised if you can't relate what you see to those four words. (Mintzberg 1973: 49)

Overall, the picture that emerges from these studies is that of a wide variation in behaviour by managers in different jobs, and also by managers in the same job. Managers, as individuals, face a complex, fragmented, varied and ill-defined set of demands, which they deal with by a range of ad hoc responses. Commenting on such studies, Cave and McKeown argue that a proper conclusion is that 'generalizations about the nature of management are dubious' (Cave and McKeown 1993: 123).

Behaviour and Context

However, before readily accepting such a conclusion, we should consider the question of *why* we might expect to discover what management is by observation of the behaviour of managers. For such attempts fall foul of the long-established recognition that all human behaviour may be viewed in fundamentally different ways: that is, any statement about the behaviour of an individual may be subject to different 'logical grammars'. In a philosophical critique of behaviourism, Hamlyn (1953: 190) pointed out that we talk of human behaviour in many different ways, at various levels of generality and with varying degrees of abstraction, from different points of view. It is important not to confuse or run together these different modes of talk.

Critics of observational studies of management (for example Carroll 1987, Burgoyne 1989, Hales, 2001) argue that these are problematic in that any observational study requires a framework of categories of behaviour into which specific observations may be located. They argue that such a framework constitutes some prior understanding of what is being intended and attempted, within a given situation, without which it is impossible to decide what particular aspects of behaviour are significant for the research purposes at hand. There have been, and no doubt will continue to be, attempts to develop frameworks for categorizing managerial behaviour capable of locating, by observation alone, specific items of behaviour by specific managers.

However, to base a category framework on what the individual (manager) attempts or intends is itself misleading. What is *socially significant*, that is to say, what is *socially consequential*, in a situation is not merely what an individual attempts or intends but how their behaviour is construed as *behaviour of a particular kind*. That is, an individual's behaviour has social significance and consequence to the extent that others attribute meaningfulness and they themselves act in a way that is related to the meaning attributed. Indeed, it is more appropriate to speak of 'acts' and 'actions' when referring to meaningful human behaviour. Harré and Secord (1972) distinguish between 'mere' movement, action by the individual and acts that are performed by the engagement in action.

> All action acquires its social meaning through being identified as a performance which, when completed, constitutes by convention an act. Acts are not to be identified either with the actions needed to perform them, nor with the movements involved in the action. There are many different ways in which the same act may be performed. All have the same meaning through their identity with the respect to the act, of which they are the performance. (Harré and Secord 1972: 11, emphasis in original)

This 'acquisition of social meaning' by 'convention' emphasizes the importance of examining the social context within which the observable activity by individual managers is *construed as* 'managing'.

Moreover, to say that certain behaviour acquires meaning by being construed as 'managing', constituted as such 'by convention', does not permit us to assume that such meaning can be taken as unitary and stable. Rather, it is likely that there will be contestation over the meaning, between the various social actors, or over time, or in different situations, or some combination of these. The fact that *certain* meanings or understandings of the nature of management have become dominant, and continue to be dominant, itself needs examination.

Examining Management in Context

The immediate context within which such construal takes place is that of *organization*. That is, we must consider not just the particular organization within which a particular individual person, *as a* manager, engages in particular forms of behaviour, but the existence of *organization* as a collective social form within which the social position of manager has social consequence. Our purpose in this volume is to examine how particular understandings of management have come to take priority over other modes of organizing collective economic and social activity: the dominance of management.

In the next chapter, therefore, we shall be examining key issues concerning the nature of this social collective form, within the societal form that has arisen in the modern era, that is, capitalism. In Chapter 4, we shall examine issues relating to the origins of management within the mode of production that developed with the rise of capitalism. Following this, in Chapter 5, we shall consider matters concerning management within the form of the state under capitalism. The subsequent five chapters will be devoted to examination of the events that unfolded in the two organizations described in Chapter 1, the Stonebridge Bus Garage Project and the Charlton Training Centre.

Chapter 3
Radical Organization Theory and Research

Which Theory of Organization?

As we have noted, modern management has arisen alongside the development of the modern form of organizing, that is *as organizations*. By the early twentieth century, the development of corporations and joint-stock companies as equivalent under law as distinctive persons was well established (Pollard 1965). Modern society may be characterized as an 'organizational society' (Presthus 1979). What were originally 'legal fictions' (Hodgson 2002) took on the character of a 'natural fact' within the dominant literature on organizations, where typical texts would talk of 'organizational behaviour' and 'behaviour in organization'. Thus organizations were treated as *entities* (Hosking, Dachler and Gergen 1995), bounded and separate from their 'environment'.

Until relatively recently, textbook presentations of organization theory, often containing the word 'behaviour' in their titles, have typically presented the history of the field as one of progressive discovery of knowledge. Their content is very similar to that of management textbooks; this is understandable as their intended audiences were typically those studying for business and management qualifications. However, there can be no doubt that the field of organization studies has changed dramatically over the past two to three decades. Introducing a collected edition of writings on the field of organization studies in the late 1990s, Clegg and Hardy (1999) point to the major changes that had taken place since the 1960s: politically, economically, socially, technologically. Since then, major changes has not only changed the 'terrain' of organization studies but had also produced new approaches. In the 1960s, according to Clegg and Hardy, it seemed that what Atkinson (1971) termed an 'orthodox consensus' was emerging, based around functionalism, 'by which we mean an approach premised on assumptions concerning the unitary and orderly nature of organizations' (Clegg and Hardy 1999: 1). Since then, a wide range of alternative approaches have developed.

In a contribution to the same volume Reed argues that whilst by the 1950s and 1960s, organizational studies texts 'bridled with self-confidence concerning their 'discipline's' intellectual identity and rationale', over the 1980s and 1990s, such self-confidence had 'drained away', replaced by 'uncertain, complex and confused expectations concerning the nature and merits of organization studies' (Reed 1999: 26). A landmark text in capturing and contributing to such major change in the field was the study by Burrell and Morgan (1979). This drew upon Kuhn's notion of paradigms, used in his analysis of the history of science and the structure of scientific revolutions (Kuhn 1970).

Burrell and Morgan attempted to locate organizational theory and analysis within a framework of 'sociological paradigms', developed from the central idea 'that all theories of organization are based upon a philosophy of science and a theory of society' (Burrell and Morgan 1979: 1). They conceive the philosophy of science upon which a particular theory may be based as being somewhere on a dimension of 'subjective' to 'objective'. At the subjective end of the dimension the phenomena under investigation are treated as products of one's mind; at the objective end the phenomena are treated as 'hard facts', 'out there'. The objective end is positivist in its epistemology whereas the subjective end is anti-positivist. Those who hold to a philosophy of science at the objective end regard human beings as being determined by external factors in the social world and therefore seek general laws of nature using similar approaches as those of the traditional 'natural scientist'. Those who hold to the subjective end see humans as having considerable self-determination and as creating, modifying and interpreting the world in unique ways.

Added to this dimension of the assumed view of science is the dimension relating to the theories of society inherent in different schools of sociology. At one end of this dimension is the 'sociology of regulation' which is 'primarily concerned to provide explanations of society in terms which emphasize its underlying unity and cohesiveness' (Burrell and Morgan 1979: 17). The elements perceived to be important are such aspects as cohesion, consensus, integration, co-operation. The underlying assumption is that certain forces are at work which maintain the status quo and prevent social order disintegrating into chaos. At the other end of this dimension is the 'sociology of radical change', which is based on the view that man is prevented by structures of society from achieving full development. This sociology focuses on the modes of domination, coercion, hostility, and conflict, particularly between forces acting towards and forces acting against radical change.

From these two dimensions Burrell and Morgan devise a framework for examination of the various schools of organizational theory. Four 'sociological paradigms' are identified: 'functionalism', 'interpretive sociology', 'radical humanism', 'radical structuralism'. The various schools of organizational theory are located on this 'map' of the four sociological paradigms. Burrell and Morgan argue that 'whilst the activity within the context of each paradigm is often considerable, inter-paradigmatic "journeys" are much rarer', continuing:

> the four paradigms are mutually exclusive. They offer alternative views of social reality, and to understand the nature of all four is to understand four different views of society. They offer different ways of seeing. ... one cannot operate in more than one paradigm at any given point in time, since in accepting the assumptions of one, we defy the assumptions of all the others. (Burrell and Morgan 1979: 24-25)

Contesting the Paradigms Approach

Burrell and Morgan use their framework of 'sociological paradigms' to review the field of organizational analysis and the sociological theories underlying the various schools of analysis. Others had also adopted the term 'paradigm', from Kuhn, to explain diversity in sociological theory, including Friedrichs (1970), Effrat (1973), and Eisenstadt and Cureleu (1976). In applying the notion to the field of organization studies, Burrell and Morgan effectively challenged the assumption in the dominant mainstream organizational behaviour literature that it was the singular correct approach.

However, Harvey criticized such efforts as 'arbitrary pigeon-holing schemes, of varying degrees of sophistication', 'personalized schematic devices', arguing that 'the labelling of such pigeon-holes as paradigms legitimizes the scheme and implies an authority it does not possess' (Harvey 1982: 87). He argues that such writers, including Burrell and Morgan, have misused Kuhn's term 'paradigm', which Kuhn used in relation to his conclusion that 'normal science' is undertaken within a prevailing theoretical framework, and that revolutions in science occur only after theories within the theoretical framework have become extended and complicated so much that they verge on self-contradiction. To use the term 'paradigm' in such a way that does not accord with this notion of the competition between one ascendant theoretical framework in difficulties and an opposing framework which eventually replaces the former, these opposing frameworks being mutually exclusive (the principle of incommensurability), is in Harvey's view non-Kuhnian.

Other writers have also debated Burrell and Morgan's analysis, especially the view that paradigms are incommensurable (Jackson and Carter 1991, Willmott 1993, Weaver and Gioia 1994). Hassard (1990) argued that a *multi-paradigm* approach might fruitfully be adopted for organizational research, applying this to a case study of the British Fire Service (Hassard 1991). Kamoche (2000) similarly used the Burrell and Morgan framework to engage in a multi-paradigmatic study of human resource management in an African context. Gioia and Pitre (1990) suggest that the boundaries between the four paradigms in the model are ill-defined and 'blurred', permeable, better considered as 'transition zones'.

Whatever the resolution, if any, that might be achieved on the debate about specific aspects of the Burrell and Morgan framework, it has clearly been very influential in promoting pluralism in organizational studies (Hassard and Pym 1990). More importantly, for the purpose of this book, it opens up the opportunity for, and warrants, consideration of modern management within an alternative paradigm to that which has dominated over the past century or more of its rise and development. In particular, we shall be mainly concerned to establish the basis of an approach to the analysis of organization and of management which, following Burrell (Burrell 1980), will be referred to as 'radical organization theory'.

Elements of Radical Organization Theory

We can identify three main elements of research undertaken within such a framework. It starts from fundamentally different assumptions about the nature of society. Such research is undertaken in the interests of those who in society are oppressed, subject to domination, exploited. It adopts a different strategy, attempting to study the social world in a way that explains the underlying structures which give rise to 'false appearances' and, in doing so, to move beyond them.

In respect of the first of these elements, we have already noted that the underlying assumptions about the nature of society on which 'functionalist' organization theory is based differ fundamentally from 'radical' organization theory. These are expressed by Burrell and Morgan (1979) in terms of 'sociology of status quo' and 'sociology of radical change'. The second element is expressed by Burrell who argues that radical organization theory is 'an approach which is radical because its primary aim is to study the structure of contemporary capitalist societies and use the knowledge thereby attained in the interests of the exploited social majority' (Burrell 1980: 90).

The third element, the nature of the research process itself, is dealt with by many writers who variously refer to the 'Marxian method', 'dialectical analysis', and 'organizational praxis' (Appelbaum 1978, Benson 1977, 1983, Heydebrand 1983, Wright 1983, Carchedi 1983). These writers see the research process as a process of knowledge 'production' which 'is like other forms of social production, shaped by its context and by the way that knowledge producers are inserted into the social world' (Benson 1983: 334). In order for these to form the elements of a cohesive paradigm, it is necessary to examine how they are integrally linked to each other.

The Nature of Society

We have seen that Burrell and Morgan locate radical organization theory within what they term the 'radical structuralist' sociological paradigm. They argue that this paradigm, although based on a 'sociology of radical change', shares with the functionalist paradigm similar ontological and epistemological approaches. Although the actual ontological and epistemological assumptions differ, radical structuralism shares with functionalism the assumption that social phenomena are best understood as 'real' that is, not dependent on individual consciousness and that knowledge of social realities is relatively independent of the individual knower. We shall focus here mainly on the other 'dimension' for analysis used in the Burrell and Morgan framework, that is, the regulation versus radical change dimension. 'Sociology of regulation' assumes society to have underlying stability, and seeks to provide explanations for such cohesiveness. In stark contrast the 'sociology of radical change' is concerned to 'find explanations for the radical change, deep-seated structural conflict, modes of domination and structural contradiction which its theorists see as characterising modern society' (Burrell and Morgan 1979: 17).

The intellectual origins of radical structuralism lie mainly in the work of Marx, directly through the writings of Marx and later avowedly Marxist writers, and indirectly through the work of Weber. The work of Weber has been variously interpreted by later theorists, and his writings on bureaucracy have been incorporated in much of functionalist organization theory. However, Salaman argues that, despite some obvious differences in their approaches to the study of large-scale organizations, they share certain interests in organizations, ask similar questions of them, and focus attention on similar processes.

> They firmly rejected the widely prevalent view that organizational structure follows from the application of neutral, apolitical priorities – such as efficiency, technology, etc. – and insisted that such concepts should be exposed for their political purposes and assumptions, and focus attention on the nature and function of organizational ideologies. (Salaman 1979: 23)

Burrell and Morgan use the phrase 'radical Weberianism' to describe the contemporary resurgence of interest in Weber's analysis of bureaucracy as an instrument of social domination, arguing that Weber 'did joust with the Marxian heritage and fought the battle on its ground, at least on occasion, and it is the product of this sort of confrontation which forms the kernel of contemporary "radical Weberianism"' (Burrell and Morgan 1979: 333).

However, it is primarily in the Marxist tradition that I shall be seeking the basis for the development of a radical critique of management. Of course Marxism is not one clearly delineated set of propositions about society. Many Marxist writers have criticized the naive assumptions made by non-Marxists about what Marx wrote and about later developments of his work. Much of this equates Marxist theory with the practice of particular states which declared a commitment to Marxism, notably the former Soviet bloc states and China. But as Swingewood (1975) points out Marx completed few of his mature projects, and his ideas underwent changes during the course of his development of them. Moreover Marx was mostly concerned not with a complete abstract theory of society but with an analysis of the emerging society in which he was living, a society which was based on a particular mode of production, namely capitalism. Noting these points however, we can sketch out what for Marx are the main elements of the theory of society he used to analyse the capitalist society which was then emerging.

First, all societies are stratified into distinct classes, which have their basis in the economic organization of society. Under capitalism the relationships between classes are simplified into one of fundamental conflict between the owners of capital, the bourgeoisie, and the workers who own no capital but are forced to sell their labour power in order to live, that is, the proletariat. Secondly society is a totality: the economic base, that is, the mode of production, is closely bound up with the 'superstructure', that is, the relations between classes and the institutions for reproducing society. Thirdly social processes are contradictory and dialectical and social change is more revolutionary than evolutionary. Fourthly, there are laws which characterize society and history; human beings themselves do create history through practical activity, but the conditions under which this is done are not totally created by them. Fifthly, class society continues as much by 'false consciousness' (ideology) as by force.

A radical approach to organizational research would need to start from these fundamental elements of Marxist theory. It would need to adopt an approach which was one of 'political economy' that is, being concerned to analyse the relationship between the capitalist mode of production, the dominant economic form, and the processes of society and the political and social institutions. It would use the concept of class as an integral part of the analysis. The role of the state and the major social institutions would be subject to critical analysis. The interrelationships of aspects of society would be drawn out and clarified, including the relationship between the labour process within the capitalist mode of production and the division and conflict between the two main classes. The part played by ideology in reproducing capitalist society would be examined. Clearly within such an approach, 'management' as an abstract, ahistorical concept has little meaning. Management as a social phenomenon of modern society has had a historical development. The historical development of capitalism and of rational-domination through bureaucracy must form the basis for understanding the existence of management in modern organizations and the part played by management in the reproduction of the dominant mode of production that is, capitalism. Moreover the concept of organization itself would need critical examination.

Much organizational theory is based on a reification of this concept, whereby organizations are presented as mystifications, phenomena which have existence in themselves not as the creations of humans. Instead the processes of exploitation and domination inherent in capitalism must be traced in the social practices in organizations. Nor can these be regarded as simple and predetermined. Marx mainly examined the labour process in the production of material commodities, but clearly advanced capitalist societies have developed a wide variety of organizations which do not produce material commodities. In particular the development of the 'welfare state' has given rise to many forms of organization which may be analysed not only in terms of the exploitation of workers in those organizations but also in the part they play in maintaining class exploitation.

Researching Organizations Radically

The strategy appropriate for radical organizational research can be considered in terms of the key concepts about the nature of social reality, and the methodology which is capable of enabling the researcher to develop an understanding and explanation of that reality. Of course these are integrally connected, for the assumed nature of social phenomena will both determine and be determined by the methodology considered to be appropriate. Thus within conventional theory and research, which Burrell and Morgan refer to as 'functionalist', the assumed 'concreteness' of 'social facts', and the empiricist–positivist methodology ('the scientific method') go hand in hand. Radical organization research has a coherent conceptual and methodological framework which will now be examined.

Burrell and Morgan (Burrell and Morgan 1979) identify four notions which are central to radical organization theory and research:

- *totality*: social relations within organizations are integrally connected with social totality;
- *structure*: configurations of social relations are relatively persistent and enduring, and exist independently of men's conscious of them;
- *contradiction*: structures are seen as posed in contradictory, antagonistic relationships one to another;
- *crisis*: contradictions within a given totality reach a point at which they cannot be contained, at which point one set of structures is replaced by another of a fundamentally different kind.

Totality and Structure

The notion of totality refers to the requirement that any aspect of society under study must be examined within the total social formation. As Benson argues, it is important to pay attention to the multiple interconnections within social phenomena: '[a]ny particular structure is always seen as part of a larger, concrete whole rather than as an isolated, abstract phenomenon' (Benson 1977: 3-4). Totality implies, according to Burrell and Morgan, 'that organizations can only be understood in terms of their place within a total context, in terms of the wider social formation within which they exist and which they reflect' (Burrell and Morgan 1979: 368).

This view of totality should not be identified with the view of system and environment adopted in much functionalist organization theory. Such conceptualization takes the organization (system) as the unit of analysis, and the relationship with the environment as a highly specific and neutral one. Adopting such an approach distracts researchers 'from any genuine sociological interest in the relationship between the nature of society and the contribution of organizations to its maintenance and development. This trivializes their analysis' (Salaman 1979: 10).

Similarly, Therborn argues that 'there remains in all these approaches a basic dichotomy between the organizational subject and its "setting" – a dichotomy which hinders deeper consideration of the processes of social reproduction and change' (Therborn 1978: 37). Radical organization theory regards the social world in processual terms – social arrangements appear fixed and permanent, but in reality are continuously being constructed and reproduced. Therborn expresses this, admittedly simplistically, in diagrammatic form (see Figure 3.1).

However, contrary to ethnomethodological and symbolic interactionist approaches which tend to regard all aspects of the social world as 'negotiated',

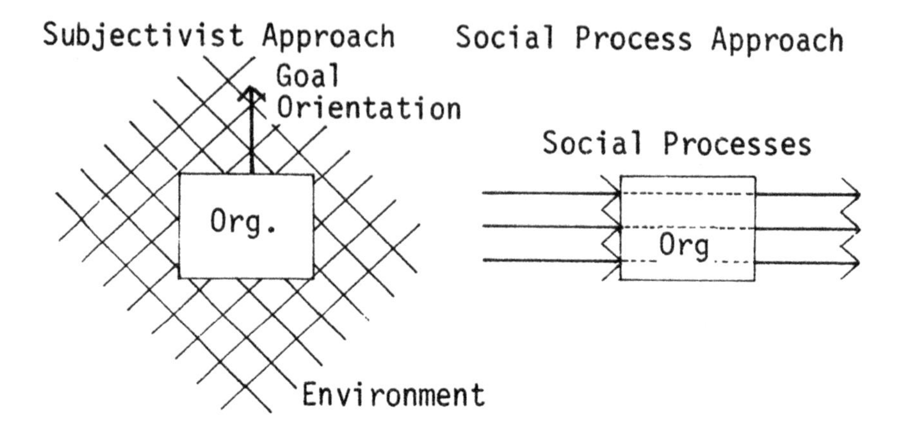

Figure 3.1 Therborn's model of Subjectivist Approach and Social Process Approach

Source: Figure reproduced from Therborn (1980), courtesy of Verso.

radical theory views such social construction/production as taking place within identifiable structures or social patterns which tend to be relatively persistent. The structures within the social world arise from deep seated generative mechanisms within the totality of the social world. Structures are constantly being produced and reproduced by people who in turn are produced and reproduced by those structures. As Marx asserts in a famous passage,

> Men make their own history, but they do not make it just as they please; they do not make it under circumstances chosen by themselves, but under circumstances directly encountered, given and transmitted from the past. (Marx 1953a: 225)

Contradiction and Crisis

The notions of contradiction and crisis are central to Marx's method for analysis of society. Change comes about, according to Marx, because of contradictions within the totality. Structures within the totality are seen as being posed in antagonistic, contradictory relationship to one another. Within the capitalist epoch, which Marx was particularly concerned to analyse, various contradictions can be identified, especially between forces of production (society's capacity to produce, as determined by existing scientific knowledge, technological equipment and the organization of collective labour) and the existing relations of production (the way that ownership and control of the means of production is organized between various social groups, particularly indicated by property relations).

> At a certain stage of their development, the material productive forces of society come into conflict with the existing relations of production, or – what is but a legal expression for the same thing – with the property relations within which they have been at work hitherto. From forms of development of the productive forces these relations turn into their fetters. Then begins an epoch of social revolution. (Marx 1953b: 329)

There are thus fundamental contradictions between capital and labour, between the increasing socialization of the forces of production and the narrowing basis of their ownership. The need for individual capitalists to pursue profit-maximising strategies contradicts the interests of the capitalist class as a whole, because of the 'law' of the falling rate of profit. Such contradictory structures are held in antagonistic relationship for most of the time. From time to time adjustments are made: crises of capital accumulation and overproduction result in periodic recessions where the value of capital is depressed, thus enabling the conditions to be restored which are favourable for capital accumulation to begin again. However such internal contradictions cannot be perpetually coped with by minor readjustment and eventually cataclysmic break up of the totality leads to a new social formation. Thus under extreme conditions the inherent contradictions in the existing totality lead to 'detotalization' and a transformation to a new totality, that is, a new set of inter-related economic structure and superstructure. Such change is not gradual and evolutionary, but occurs 'suddenly, catastrophically, as crises through the medium of revolutionary activity and are not the result of slow evolutionary build-ups and drawn out transformations' (Burrell 1980: 97).

Of course, this is not to say that other non-critical change does not occur. Marx's analysis of the labour process within the capitalist mode of production is a clear description of a process of change, from 'cooperation' where formerly disparate workers are brought together under one roof, through 'manufacture' with a complete reorganization of the division of labour, to 'large-scale industry' with production becoming based on mechanization and the worker a 'mere appendage'. Such change, however, takes place within the existing totality, and can only be understood within an understanding of the totality and the contradictions which determine such change.

Dialectical Method

Marxist theory is often described as 'dialectical'. However this term is by no means solely used by Marxist writers, and has been a key concept in much of Western philosophy dating back to Plato. Williams points out that the original use of the word 'dialectics' in English from the fourteenth century was in a sense which we would now call 'logic', referring to the art of formal reasoning, through discussion and debate (Williams 1976). Such reasoning and debate seeks to reconcile two contradictory positions, and forms part of the method adopted by such philosophers as Pascal, Kant and, most supremely, Hegel. However the application of dialectics in socio-economic analysis is mostly associated with Marx and Engels. Marx never wrote a systematic treatise on dialectics, although his works are examples of dialectical analysis. Engels attempted to provide a popular and systematic rendering of dialectics in 'Anti-Duhring' in which he stated that dialectics 'is nothing more than the science of the general laws of motion and development of nature, human society and thought' (Engels 1955: 194)

The concept of dialectics is thus concerned with process of change within a totality through contradictions generating changes which incorporate the contradictory elements. It is an approach, according to Sherman, 'that visualises the world as an interconnected totality undergoing minor and major changes due to internal conflicts of opposing forces' (Sherman 1976: 57).

Various interpretations have been made of this, partly because of ambiguities in Engel's formalistic rendering and differences between Engel's description and the actual usage by Marx. Some Marxist writers have interpreted dialectics in highly rigorous fashion, seeing it as a description of natural history as well as human history. Others have interpreted dialectics to be a form of logic, which is superior to 'formal' that is, Aristotelian logic. Sherman regards such writers as 'vulgar Marxists' and argues that dialectics as used by Marx should be regarded as a method for the social sciences, and that the criticisms made by anti-Marxist writers using the vulgar Marxist view of dialectics thus do not apply. Similarly Swingewood states that Marx, when he wrote *Capital*, had developed a distinct methodology for the analysis of social, economic and political structures' (Swingewood 1975: 12). Sherman argues that dialectics should be subject to the test of good methodology, which is not whether it is true to any facts, but whether it is 'useful', that is, the extent to which it

> (1) directs us to choose the most important problems, (2) directs us to select the facts that are most relevant for solving those problems, and (3) helps provide a framework for interpreting the facts so as to solve the problems, formulate new theories, and lead to better social practice. (Sherman 1976: 61)

Dialectical Analysis of Organization

Within recent work on organizational analysis several writers have attempted to make explicit a dialectical approach. Although there are certain differences between the various formulations made it is possible to outline the key elements of such an analysis. Drawing on the outlines of the Marxist analysis described, the dialectical approach seeks to elicit the contradictions implicit in any given totality and to explain the processes by which crisis arises and results in fundamentally different structures. However the results of crisis is not the replacement of one of the antagonistic structures by the other, but a totally new configuration of structures. Heydebrand (1983) and Benson (1983) refer to this process as 'praxis'. For Benson, this involves

> the achievement of a reasoned basis for emancipatory action – action that removes unnecessary constraints on the development of human societies and opens new possibilities where human productive activity can more freely realize human potential for self-organization. (Benson 1983: 335)

The notions of contradiction and crisis are not separate from and additional to the notions of totality and structure but arise logically from those notions. Benson (1977) sees contradictions arising through a given form of social organization developing in accordance with its inherent tendencies which generate practices that threaten its own essential character or reach beyond its limits. He goes on to point out that the structure of an organization generates bases for organizing action in the form of interests and power bases which potentially challenge the existing order. Carchedi (1983) argues for a class analysis approach to organizations which stresses the formation of social classes by the grouping of people according to the relations of production. This is based on the primacy of the transformation of material reality, that is, production, in the formation of society. Carchedi argues that under capitalism these relations of production are antagonistic and asymmetric. Antagonism results from the tie of mutual existential dependence of two antagonistic poles, for example owner–non-owner, exploiter–exploited, non-labourer–labourer. They are asymmetrical because one aspect is the principal or determinant one, the other the secondary or determined one.

In order to identify such contradictions it is necessary to identify the determinant elements and the determined elements within given social relations, recognizing that these exist within a social totality and not in a social vacuum. This is both a logical/theoretical process and an empirical/historical process, and dialectical analysis must combine both of these. Benson (1983) argues that this involves investigating social arrangements *concretely* rather than abstractly, locating events in their total contexts rather than abstracting them from their contexts. So, 'a prominent feature of the analysis must be the location of observations within total social formations i.e. in relation to the core tendencies of the social formation' (Benson 1983: 333). Such a 'totalizing move' involves dealing with the relations between what are the 'essential, core tendencies of a social formation' and the contingent, historically specific forms taken; this requires

> requires getting past the confusing array of factual observations to a conceptual model of the social formation that sorts out the essential from the nonessential components, locates events within strata or levels, and identifies the main developmental tendencies. (Benson 1983: 334)

Similarly Carchedi argues that, following Marx, we reach the 'rich totality of many determinations and relations' not by abstracting the historical essence but

> on the contrary, by focusing, by condensing it in what is specific and historically determined. … Logical and historical analysis thus complement each other.

> The former type of analysis provides the scheme for interpreting the relation between different instances; i.e. it provides the concept of determination, but the adjudication of the determinant, dominant and so on to those instances is a question for historical analysis. (Carchedi 1983: 354-355)

Conclusions for Research

From the above certain conclusions may be drawn regarding the strategy for researching management and management education and training. 'Management' as an abstract, ahistorical concept is relatively meaningless. Management as a social phenomenon of modern society has had a historical development. The historical development of capitalism and of rational-domination through bureaucracy must form the basis for understanding the existence of management in modern organizations and the part played by management in the reproduction of the dominant mode of production, that is, capitalism.

We shall examine three related approaches to this: labour process theory, class analysis, and state theory. Labour process theory is the body of theory and empirical research which seeks to understand and explain the specific dynamics of the capitalist mode of production within the actual productive process. Although this formed a central section of the first volume of Marx's *Capital* it was relatively neglected in post-Marxist literature. It has again begun to be a major area of exploration in recent years, mainly as a result of Braverman's *Labor and Monopoly Capital* (1974). Braverman's thesis is that capitalism has continued in its inexorable development into its latterday form of monopoly through the formulation of earlier forms of subordination of labour to capital into 'Scientific Management' or 'Taylorism'. This thesis has been subjected to criticism and refinement but the main thrust of the research has continued. The areas now being examined within this field include the attempts by management, as agents of capital, to increase the amount and rate of the generation of relative surplus-value through strategies of direct control, deskilling, manufacture of consent, legitimization of control, overcoming resistance, racial and gender divisions of labour.

A radical approach provides a serious theoretical and empirical challenge to the 'orthodox' view of management as a rational-technical function. Related to labour process theory is the analysis of the class position of managers and other professional, 'middle class' workers. This approach examines the class position, not as functionalist theorists would merely in terms of socio-economic status, but specifically from the perspective of the class relations of production under capitalism. Classical Marxist theory assumes two main antagonistic classes under capitalism, bourgeoisie (capital) and proletariat (labour). The third class, the petit-bourgeoisie consisting of small tradesmen, independent shopkeepers, independent farmers, etc., were regarded as subject to a continuing process of proletarianization, eventually decaying into the proletariat proper. The development of capitalism has however seen a rise in a social group central to the capitalist mode of production, which on the face of it cannot be described as bourgeois or proletariat. The members of this group do not own the resources for production yet act on behalf of the interests of capital.

Debate on the class position of this group has opened up a wider debate on the nature of classes and the processes of class formation and reproduction (e.g. see Walker 1979). Within this debate the role of ideology is of significant importance. In terms of the analysis of management education and training the approach of Child (1968) is illuminating. As we noted earlier, Child pointed out two facets of management thought, the technical content dealing with the application of 'rational, scientific' approaches to work organization, and also the legitimatory content intended to gain social approval for management activities and indeed management itself.

State theory is an area of research and analysis which has played very little part in the study of management and in the practice of management education and training. Significantly, however the role of management, management ideology, and management education and training practice have played an important role, certainly in recent years, in state processes. The authoritarian style of Conservatism of the Government during the 1980s (the time period of the two organizations that are examined in this book) was marked by calls for greater application of 'sound management' principles in a wide range of state activities: nationalized industries, health and welfare services, policing, defence and so on. Indeed one of the mottos (or slogans) of the Conservative Government's industrial relations strategy was 'giving back to management the right to manage'. This emphasizes the need to examine management and management education and training within the context of the social totality at the level of the state. Therborn's paper and subsequent book, both entitled *What Does the Ruling Class Do When it Rules?* provide useful approaches for such analysis (Therborn 1970, 1978). He argues that, in order to reproduce society, positions within the given social structure have to be reproduced as do the individuals who occupy those positions. The former aspect of reproduction, since it requires the (re)production of compatibility between different strata of the social structure, requires the reproduction of not only capitalist enterprise but also a compatible capitalist state.

The latter aspect of reproduction requires that

> new generations – and the given individuals year in and year out – have to be trained to occupy the given positions, to be qualified or subjected to fulfill adequately the tasks provided by the social structure. (Therborn 1970: 15)

The mechanisms of such reproduction obviously include ideology. However Therborn also points to economic constraint and political mechanisms, both administrative and repressive. The resurgence of management's claim to the 'right to manage', particularly evident in industrial relations contexts (Storey 1982) can be examined in the light of such mechanisms. Of greater relevance for the events described in the field studies is the nature of the interventions of the state in the organization and management of community initiatives. These interventions will be shown to be of critical importance, subverting any potential for radical change and thereby supporting the interests of capital in maintaining the capitalist characteristics of the state.

We shall examine in more depth these three areas over the next three chapters.

Chapter 4
Management and the Labour Process

Management or Capital?

By the mid-twentieth century, Peter Drucker, the foremost writer on management was able to state that

> [w]e no longer talk of 'capital' and 'labour'; we talk of 'management' and 'labour'. The 'responsibilities of capital' have disappeared from our vocabulary together with the 'rights of capital'; instead, we hear of the responsibilities of management', and (a singularly hapless phrase) of the prerogatives of management. (Drucker 1968: 13)[1]

Drucker's statement may be accepted as true within 'orthodox' circles of managers, management educators and management theorists. However within radical and critical theories of organizations and management the statement and Drucker's views would be treated as an oversimplification of the origins and rise of management within capitalism. One of the key areas for examination, especially since Braverman's *Labor and Monopoly Capital* (Braverman 1974), is the place of management in the labour process under the capitalist mode of production. Because Marxist theory provides a coherent and radically alternative analysis of dominant modes of organizing productive activity from the analysis on which 'orthodox' management theory and education is based, Marxist theory(-ies) of the labour process should figure large in any attempt to develop a radical-alternative management theory.

Marx's Analysis of the Capitalist Labour Process

A large part of Marx's life and work was devoted to the analysis of the major economic, social and political changes taking place in the industrialized and industrializing nations in the latter part of the nineteenth century. A key theme within his analysis is that capitalism is based on, and requires for its continuation, major change in the way that work is organized, the labour process. Marx provides a framework for examining the labour process under capitalism in the first volume of *Capital*. 'Labour', says Marx

1 Originally published in 1955.

> is, in the first place, a process in which both man and Nature participate, and in which man of his own accord starts, regulates, and controls the material re-actions between himself and Nature. (Marx 1954: 173)

He stresses that in talking about human labour he is dealing with purposeful activity, not the instinctive behaviour of, for example, spiders in 'weaving' webs or bees in 'constructing' hives. He identifies three elements of the labour process: the personal activity of humans, that is, work itself; the object on which the work is performed; and the instruments of work, that is, tools or more complex technology. Marx describes how these three elements combine to produce the end results of productive labour:

> In the labour process, therefore, man's activity, with the help of the instruments of labour, effects an alteration, designed from the commencement, in the material worked upon. The process disappears in the product; the latter is a use-value, Nature's material adapted by a change of form to the wants of man. (Marx 1954: 176)

This simple and neat classification of the elements of the labour process is further elaborated by Marx in order to make clear the basis of the process. Thus he shows that the objects on which work is performed can be either spontaneously provided by Nature (e.g. fish caught in the water, timber felled in virgin forest, and ores extracted from their veins) or themselves 'filtered through previous labour', which he calls 'raw material'. He then goes on to argue that, except in extractive industries, all branches of industry manipulate raw material, which are products of *previous* labour. This is even true in agriculture where animals and plants are themselves the products of generations of development under human superintendence and through human labour. In a similar way Marx shows that the instruments of labour, except in basic forms such as bodily limbs and the earth and Nature itself, are products of previous labour. Instruments include those which are not directly involved in the labour process but without which the process would not take place: premises, roads and so on. Marx's concern is, however, more than mere description of the elements of the productive labour process in whatever economic epoch. His aim is to explain the particular characteristics of the labour process *under capitalism*.

Two features stand out as characteristic of the capitalist mode of production. First, the labourer works under the control of the capitalist; secondly the capitalist, not the labourer, owns the labour power of the labourer, because it is a commodity which he has purchased. The purpose of the capitalist is to create surplus value, that is, to create value in excess of the use-value created by the elements of the labour process. To do this, *surplus value*, value in excess of the exact equivalent of the labour power used must be created. This process of creating surplus value, valorization, becomes the distinctive feature of the capitalist process. In the early phase of capitalism the capitalist was obliged to accept labour-power much as it existed under pre-capitalist forms of production. Thus the skills and methods of working were little changed during this early phase. However, Marx argues, the fact that the capitalist as the purchaser of labour power now controlled the use of that labour power and the products created by it, enabled capital to pursue the objective of subordinating labour, thus enabling valorization to increase.

This subordination of labour consisted of two parts: first, the *formal subordination of labour*, in which although labour becomes intensive, longer and under the supervisory eye of the capitalist, the character of the mode of working is not affected; followed by the *real subordination of labour*, brought about by the development of productive forces, introducing and applying science, machinery and the large-scale approaches to production. In accomplishing this the mode of production goes through three phases, according to Marx. The first phase he called 'cooperation' during which workers are brought together under one roof and under the control of the capitalist. Next comes the period of *manufacture* under which the division of labour takes place, either by bringing together previously separate crafts or by splitting up work into partial operations. Finally, real subordination of labour is achieved fully within large-scale industry, in which the mechanization of work takes over: '[a]long with the tool the skill of the workman in handling it passes over to the machine ... In the factory we have a lifeless mechanism independent of the workman, who becomes its mere appendage' (Marx 1954: 396, 398). Continuing, he argues that every form of capitalist production,

> inso far as it is not only a labour process, but also a process of creating surplus value, has this in common, that it is not only the workman that employs the instruments of labour, but the instruments of labour that employ the workman. But it is only in the factory system that this inversion for the first time acquires technical and palpable reality. (Marx 1954: 398-399)

The advance of machinery-based production thus results in deskilling, the reduction and destruction of handicrafts: put starkly, 'the instruments of labour strike down the labourer' (Marx 1954: 407). Marx quotes Ure, writing some years previously:

> Whenever a process requires peculiar dexterity and steadiness of hand, it is withdrawn, as soon as possible, from the cunning workman, who is prone to irregularities of many kinds, and is placed in charge of a peculiar mechanism, so self-regulating that a child could superintend it. (Ure 1835: 19-20, quoted in Marx 1954: 407)

And so Marx continues, referring to authorities and evidence of various kinds, in his exposition of the advance of capitalism and of accumulation of surplus value through the real subordination of labour.

Braverman and Labour Process Theory

Despite, or perhaps because of, the detailed analysis provided by Marx, little attention was paid by theorists and researchers to the labour process under capitalism for nearly a century after Marx's death. There were various attempts at 'humanizing' work within capitalism, which thus implicitly assumed that there was no essential conflict between the capitalist mode of production and the aim of such attempts, no recognition that subordination of labour was central to the labour process under capitalism. The seminal work that challenged such a view was Braverman's *Labor and Monopoly Capital*, published in 1974. Thompson (1983) refers to 'the explosion of interest in the labour process that followed in its wake'. Storey (1982) refers to it as a 'catalyst for much contemporary debate within the marxist and neo-marxist corpus'. Sweezy, in his foreword to Braverman's book, refers to the introduction to his own book, co-authored with Baran (Baran and Sweezy 1968), where they stated that they were conscious of neglecting the labour process, a subject central to Marx's study of capitalism. He continues that '[n]ow at last, in Harry Braverman's work ... we have a serious, and in my judgement solidly successful, effort to fill a large part of this gap' (in Braverman 1974: ix).

Braverman rejects the 'conventional' view that there has been an upgrading of work in recent times, that the proportion of the working population involved in manufacturing and related industries was shrinking in favour of 'white collar work requiring higher levels of education and training to perform work with greater mental effort'. He stresses the central feature of capitalism, the creation and accumulation of profit rather than the satisfaction of human needs. He argues that capital needs to realize the potential it has in the purchase of labour power by transforming it into labour power under its own control – the real subordination of labour in Marx's terms (although not explicitly referred to as such by Braverman). In the struggle to bring this about modern management is developed, devising the means to impose the will of capital within the new social relations of production. The division of labour and systematic reduction and destruction of craft and skill becomes generalized as the capitalist mode of production rapidly and seemingly inexorably gains dominance in all spheres of production.

Whilst recognizing that in previous epochs the management of activity and control of large bodies of workers took place, Braverman argues that a qualitative change took place under capitalism.

> The capitalist, however, working with hired labor, which represents a cost for every non-producing hour, in a setting of rapidly revolutionizing technology to which his own efforts perforce contributed, and goaded by the need to show a surplus and accumulate capital, brought into being a wholly new art of management, which even in its early manifestations was far more complete, self-conscious, painstaking and calculating than anything that had gone before. (Braverman 1974: 65)

The 'new art of management' finds full expression, according to Braverman, in Taylor's 'Scientific Management', 'Taylorism'.

Braverman is careful to reject Taylorism as a true science, seeing it as 'scientific' only in the sense that it attempts to apply the methods of science within the context of the growth of capitalism and in the service of capitalism by furthering the control of labour. It starts, he argues, from the capitalist point of view, and 'enters the workplace not as the representative of science, but as the representative of management masquerading in the trappings of science' (Braverman 1974: 86). Braverman also rejects the idea that Taylorism has been superseded by later schools of industrial psychology and human relations because these take the on-going production process as given and seek to adjust the worker to this process. He refers to the practitioners of 'human relations' and 'industrial psychology' as 'the maintenance crew for the human machinery' (Braverman 1974: 87).

Braverman's Analysis of the Principles of Taylorism

Braverman examines Taylor's own writings in order to draw out the key principles of Taylorism. He refers to the first principle as *the dissociation of the labour process from the skills of the* workers, rendering the labour process independent of craft, tradition, and the workers' knowledge. 'Henceforth', argues Braverman, 'it is to depend not at all upon the abilities of the worker, but entirely upon the practices of management' (Braverman 1974: 113).

The second principle is *the separation of conception from execution*, the concentration of knowledge as the exclusive preserve of management. Workers not only lose control over their instruments of production, but 'they must now lose control over their own labor and the manner of its performance' (Braverman 1974: 116). This is necessary for capital, because if workers were in a position to guide their execution of their labour on their own conception, it would not be possible to impose the working pace and efficiency of method desired by capital. The third principle is the *use of this monopoly over knowledge to control each step of the labour process and its mode of execution.*

From this analysis of Taylorism as the method by which labour is subordinated under monopoly capitalism, Braverman goes on to show how the approach is generalized throughout industry and into the areas of 'mental work'. He subjects the work of clerical workers to analysis to show how Taylorism has taken root there also resulting in the homogenization of work and the conditions based on mechanization of office work. This is true not only of clerical operative workers but 'middle layers of workers': junior and middle managers, professional workers, and so on. Braverman posits this as evidence supporting Marx's prediction of the proletarianization of the middle classes.

Thus within this analysis, management has arisen within the development of capitalism into its monopoly stage, through capital's incessant drive to maximize profit and accumulate more capital by extracting more surplus value more effectively, from labour. Management is essentially about the deskilling of labour, separating conception from execution. It is thus not a neutral process but an aspect of the social relations of production under capitalism.

Other Analyses of the Origins of Management

Other writers have produced analyses similar to Braverman's in respect of the argument that it is the drive to generate profit which leads to the development of management as a separate function within productive enterprise. However the centrality of deskilling as the strategy for accomplishing this has been questioned either implicitly, by the analysis of other aspects of the action taken by capitalists, or explicitly by directly challenging Braverman's thesis.

Stone (1974) examines the historical development of job structures in the steel industry in USA during the period between 1890 and 1920. She also concludes that there was a deliberate strategy on the part of capitalism to destroy the existing labour process, by destroying the unions in order to introduce new technology, then introducing various schemes to undermine any solidarity and cohesion among the workforce. The schemes included wage incentive and piece work systems, new job hierarchies with competitive promotion, and welfare programmes. New divisions of labour were established with the foremen, who had previously only had authority over the unskilled workers, taking on wider authority within the new work process, and managers being recruited from outside. Various training programmes were established to 're-educate' the foremen, and to train managers recruited from outside, from technical colleges often funded by employers, in techniques of steel making.

Clawson (1980a, 1980b) examines the development of bureaucratic organization in industry, arguing that bureaucracy, meaning hierarchical management control over the work process through the separation of design and planning from actual production, was introduced by capitalists to destroy craft control. This was not because bureaucracy is inherently superior to the craft system in terms of efficiency, innovativeness, or its ability to develop mass-production methods. Instead capitalists introduced bureaucracy in order to 'be sure that the labourer does as much work as possible in one day' (Clawson 1980b: 14).

In other words, bureaucracy enables capital to realize the potential in the exchange of labour power for a wage to extract the maximum possible surplus value, to maximize profit. Clawson analyses the historical development of supervision from outside contracting, to inside contracting (both non-bureaucratic forms of organizing work), to bureaucratic organization. Under the craft system workers planned and controlled day-to-day production methods. Various cumulative management changes began to break down such control by workers, but it was the ideas promulgated by Taylor, argues Clawson, which provided for the comprehensive replacement of the craft system.

> Frederick Taylor's genius was twofold: the recognition that a system of divided control existed and would have to be abolished if capital was to dominate, and, even more remarkable, the creation and introduction of a system that provided a basis for such domination. ... Taylor introduced bureaucratic management specifically to increase control by capital, to allow capitalists full control of production. (Clawson 1980a: 255)

Stone's analysis was produced in the same year as Braverman's book, and Clawson states that his argument is compatible with, though not identical to, Braverman's.

However it is Braverman's deskilling thesis which has become the reference point from which the 'labour process debate' has ensued. Within this debate the criticisms have often been combined with analyses which extend the underlying real subordination thesis. The main arguments are (1) that capital's advance has not been straightforward and inexorable, but has been constrained by resistance by labour; (2) that deskilling is only one strategy adopted by capital and that other strategies have been adopted; (3) that wider societal processes and changes have had considerable effect on the nature of the changes in the labour process under capitalism.

A number of writers have pointed out that Braverman seems to present a highly deterministic analysis whereby he assigns only a limited role to the working class in resisting Taylorism, 'his rendering of the working class as passive, inert, living 'in accordance with the forces which act upon it' (Wood 1982: 15). Storey (1982) documents a range of forms and instances of resistance to Taylorism and discusses other factors in the development of managerial control. He argues that Braverman's neglect of worker resistance, and his exclusive focus on Taylorism and deskilling, indicate both empirical shortcomings and a key theoretical weakness.

There have, of course, been many examples of workers resisting the introduction of the managerial approaches pioneered by Taylor. Indeed much of Braverman's analysis is illustrated with quotations from Taylor's evidence to the House of Representatives' Special Committee, established in 1911 to investigate the causes of labour troubles at a US government arsenal when Taylor's approach was introduced. Some examples of worker resistance have been in the form of organized trade union action; others have been less organized, sometimes on a mass scale, sometimes on a small scale. Sherman and Wood (1979) recount the unattributed story of a millionaire eventually discovering that the hitherto inexplicable rattle in his new Cadillac is a beer can in the door compartment with a note attached saying 'So you finally found it, you rich son of a bitch'. Clawson defends Braverman in arguing that he had recognized that the activities of workers is a crucial area for analysis, and that his critics have only pointed to the problem, not helped to resolve it. However he stresses that

> workers' struggles should not be viewed as simply a 'response' to Taylorism (it is more nearly the other way around), nor as 'resistance' to capital's offensive.

> Workers' activities were not derivative from what capital did: they fundamentally shaped what happened. (Clawson 1980a: 33)

Other Managerial Strategies

The criticism of Braverman's view of Taylorism as *the* managerial strategy under capitalism has been levelled most notably by Friedman (1977b, 1977a) and Edwards (1979). Friedman argues that the problem faced by capitalists is to exercise authority over labour in order to realize the potential for exploitation. Legal authority, and the sanction of the legal right to dismiss recalcitrant workers, is insufficient because there is no guarantee that replacements would be any better. It is necessary to distinguish between capitalists' attempts to realize more of the potential surplus value, from attempts to retain the planned or expected level of surplus value. Two major control strategies adopted by employers can be identified: Direct Control and Responsible Autonomy. Whereas the strategy of Direct Control seeks to limit the scope for varying labour power using threats, close supervision and minimization of the responsibility of the worker, Responsible Autonomy 'attempts to harness the adaptability of labour power by giving workers leeway and encouraging them to adapt to changing situations in a manner beneficial to the firm' (Friedman 1977a: 78).

Friedman goes on to argue that both strategies have characterized capitalism throughout its development, but initially the Responsible Autonomy strategy was limited to privileged workers. However the increased size and complexity of firms with the advent of monopoly capitalism, and the rise of organized worker resistance, the Relative Autonomy strategy has been more widely applied. This has occurred as a result of the development of managerial theories, and the opportunity provided by monopoly power to experiment with new managerial techniques. The Relative Autonomy strategy provides for workers to have significant control over their work, but within the confines of the capitalist mode of production and the pressure to accumulate. Countering worker resistance has to be measured in terms of cost, and if a particular strategy such as Direct Control is too costly other strategies will be tried. Friedman thus argues that capitalists have a choice in respect of strategy. Similarly Edwards distinguishes between varieties of managerial strategy, arguing that capital never fully gains control over the labour process, and thus there is a 'contested terrain'. The response by labour to the strategy adopted by capital at a particular stage leads to the development of a new strategy.

Edwards (1979) distinguishes three main types of managerial strategy: simple control, technical control, and bureaucratic control. The directly personal, despotic form of control adopted in the stage of competitive capitalism resulted in intensified class struggle, leading to the development of technical control 'embedded' in the technology and work design, as seen in the assembly line and other mechanized production processes. However further struggles lead to the development of bureaucratic control, which is exercised through 'the social and organizational structure of the firm and is built into job categories, work rules, promotion procedures, discipline, wage scales, definitions of responsibilities, and the like' (Edwards 1979: 131).

Littler (1982) argues that more complex typologies are required to analyse employer strategies, which are frequently composite, mixed and even contradictory. He distinguishes three levels of structuration of work organization:

- the level of division of labour and technology;
- the level of the formal authority structure;
- the level of the wider framework of the capital/labour relationship, arising from the relation of job positions to the labour market.

Although there is a tendency for these three levels to influence each other, it is not possible, argues Littler, to collapse one level into another. Thus bureaucracy (a form of authority structure) and deskilling (a form of division of labour) do not necessarily go together. Moreover the notion of bureaucracy can be separated into two concepts: bureaucratization of structures of control; and bureaucratization of the employment relations. Littler agrees with Crozier (1964), that employers and employees have opposing interests in terms of these two aspects of bureaucracy.

> For employers and managers the preferred position is to situate workers within a tight set of rules and constraints, whilst retaining the freedom to add and subtract tasks to existing jobs, to promote 'blue-eyed' boys etc. For workers the preferred situation is strictly-controlled task allocation and promotion by seniority, combined with the freedom over work-pace and work-methods. (Littler 1982: 44)

Littler goes on to argue that the most crucial dimension of the employment relationship is that of dependence, which is determined by two broad factors: alternative sources of need satisfaction, and the capacity of workers to organize. The first of these relates primarily to availability of alternative employment, so that job security and clearly established career structures promote dependency on the part of workers. This is especially so where there is no external labour market. However the availability of provision for other needs such as health and welfare facilities will also affect worker dependency, according to the degree to which such facilities are closely associated with the employer.

Among the factors discussed concerning the ability of the workers to organize, Littler points to the fact that capitalist expansion continually creates re-occurring division among the working class, based on job-rights and job-security. As new workers enter the work process they come into conflict with those already established. Using such 'conceptual clarification' Littler undertakes a cross-cultural analysis of the labour process, in Britain, Japan, and the USA. He concludes that in the transition from indirect employment and control (through internal contracting) to direct employment and control:

> The common capitalist problem ... was the transference of the loyalty and accepted subordination from the traditional work group to a wider collectivity and a larger social frame. Bureaucratisation was not just a technical process associated with increasing size and complexity, but a sociological transformation. (Littler 1982: 192)

Finally, in this section, we should note the work of Burawoy (1979) who argues that there has been a qualitative change in capitalist strategy from coercion to control through consensual hegemony. Worker resistance is undermined and incorporated through various ways. Workers' own workplace practices, 'games', concerned with immediate rewards such as relief from boredom, reduction of fatigue and so on, themselves serve to legitimate capitalist production: '[o]ne cannot play a game and question the rules at the same time; consent to rules becomes consent to capitalist production' (Burawoy 1979: 92). The internalization within the company of competitive individualism of the external labour market, coupled with the rise of an 'internal state' in which the individual worker becomes an 'industrial citizen', with contractual rights and responsibilities, become key features in the 'manufacture of consent' at the point of production.

Other Societal Processes

Class Struggle

Thus the arguments of Friedman, Edwards and Littler point to a widening area of concern, dealing not with a limited problematic of control, but with a problematic which recognizes legitimization and consent as key elements. However the focus on the labour process itself may tend to draw attention away from other significant factors, and this tendency has been criticized by a number of writers. Aronowitz (1978) uses the term 'capital-logic' to refer to the doctrine of the subsumption of labour, science, and technology under capital, found within much contemporary Marxist literature. This doctrine has three strands: the degradation thesis (Braverman), the notion of one-dimensionality (Marcuse), and the new functions of the state in capitalist society (Miliband and Poulantzas). He argues that this aspect of Marx's work was most developed, and that the 'oppositional' side was underdeveloped theoretically by Marx and his followers. Instead the logic of subsumption should be regarded as a tendency:

> [T]he configuration of capital – including the social organization of labor, the application of machine technologies to the production process, the production of ideology and culture (and therefore consciousness) – cannot be deduced from social 'scientific' formulae according to which the entire social world appears to be a function of capital accumulation. Instead, I argue for the relative autonomy

> of labor, culture, and consciousness within the broad framework of Marxist theory of capitalist development. (Aronowitz 1978: 129)

Class struggle, argues Aronowitz, takes place not only at the point of production but also at the ideological level. Referring to Gramsci's concept of bourgeois ideology as the process by which capital secures the 'spontaneous consent' of the masses to its 'general direction of social life', Aronwitz argues that this consent must be won through the mediation of intellectuals 'who struggle for moral and intellectual leadership of society. Their understanding, of course, is far from that of conscious agents of the bourgeoisie. Rather, capital wins their consent by persuading them that intellectual life is free of the rule of society, much less capital' (Aronowitz 1978: 145).

Under certain conditions such hegemony may fail, because the consent of either the intellectuals or the masses cannot be won. The Brighton Labour Process Group argue that

> a study of the fundamental structure of the capitalist labour process necessarily abstracts from the development of the capitalist mode of production itself, and in particular abstracts from the phases of accumulation through which the labour process is connected to determinations beyond those of its general form. (Brighton Labour Process Group 1977: 23)

Similarly to Aronowitz they argue that the concrete structure of the labour process cannot be deduced from the abstract laws of capital accumulation because the production relations of capitalism can only be reproduced *as a totality* of social relations. This emphasizes the need to 'elaborate the links between changes in the capitalist labour process and changes in class composition, in political structures, in the role of the capitalist state (in education as much as the economy) and in interstate relations' (Brighton Labour Process Group 1977: 23). They point to the fact that within a capitalist social formation there are many labour processes which do not, or do not *directly*, fall under the impulse for surplus value accumulation, in particular housework and the variety of labour processes taking place under the state's command.

Education

One major area of state activity is of course education, and Sarup (1982), whilst recognizing criticisms in Braverman's work, considers it significant in indicating the need to develop an analysis of the labour process within education. He offers brief suggestions on some of the areas which need to be researched further:

- the division between mental/manual labour in the educational system and within schools;
- the managerial practices in educational institutions that parallel those in industry;

- the fragmentation and atomization of the teaching labour force.

Frith (1980) argues that one of the effects of the rise of the Manpower Services Commission (the UK state agency dealing with employment affairs during the period from 1974 to 1988) had been to dissolve the institutional separation between the responsibility which the state has for education and the responsibility which industry has for training its workforce. Young people, he argues, pose particular problems of control as the crisis in capitalism results in high unemployment:

- they are not as subject to economic control, 'dull compulsion', as workers with long term leisure or family commitments;
- this had not mattered much before because there were occupations wherein other forms of labour control were applied;
- they are believed to lack the social values which make controls tolerable and accepted;
- they lack the opportunities to develop craftsmanship and job pride because of the deskilling of occupations where apprenticeships have traditionally been used.

The various Manpower Services Commission (MSC) programmes and the calls for education to be relevant to the needs of industry are, according to Frith, attempts to deal with these problems. Similar arguments have been made in respect of initiatives by the MSC in secondary schooling, mainly the Technical and Vocational Education Initiative (Bates et al. 1984, Chitty 1986).

Implications

In examining the nature of management then, the labour process under capitalism must be a key starting point. Modern management has arisen with the development of commodity production controlled by capital in order to generate surplus value which may potentially be obtained through the purchase of labour power. In order to realize this potential capitalists have adopted various strategies to control the labour process, through 'Scientific Management' techniques and through bureaucratization of work. The actual strategies adopted at various times have been influenced not only by capital's compulsion to maintain and increase accumulation (or go under) but also by the resistance of labour. However the tendencies within the capitalist labour process are subject to external influences which potentially mediate or constrain the development of hegemonic control by capital. So management as a body of theory and values, emphasizing rational-technical thought and practices applied in a politically-neutral manner in the interests of 'progress' for society as a whole, has developed beyond the immediate application in capitalist enterprise.

Child (1969) analyses the development of management thought in Britain and concludes that there were two strands: purely technical prescriptions for organizing and administering economic resources, and legitimatory elements aimed at achieving social acceptance. What we must remember is that ideological dominance serves as much to prevent the development of alternative ideas as to gain acceptance of the dominant set of ideas. Thus capitalist management theory and practice need not obtain and retain the willing consent of labour; it is enough under 'normal' (i.e. non-revolutionary) circumstances that no cohesive alternative is developed.

Ideology, argues Aronowitz, drawing upon Althusser, is not an abstract ideal but 'a set of material practices through which people live their experience of capitalist social relations' (Aronowitz 1978: 131). Thus the experience which people have of the wide variety of institutions includes experience of managerial practices which are supported by technocratic ideology.

Reconsiderations of Labour Process Theory

The above discussion has focused on the debate about the labour process in the years shortly after publication of Braverman's book. Subsequent academic debate has further challenged his analysis, on conceptual and theoretical grounds and on the basis of empirical investigations. In particular, other strategies adopted by management for control have been explored. One key area noted is that of gender (Grieco and Whipp 1985, Pearson 1986, Bradley 1986), whereby management deploys cultural notions of femininity and passivity, and exploit the marginality of women to the labour market because of their family roles, in order to gain and maintain control over labour.

It is clear that managerial control strategies are not fixed and certainly vary more widely than Braverman recognized. These have been explored through a range of studies and the field of labour process analysis has taken a clear place in the curriculum for the sociology of work and of employment, and within critically-oriented organization studies (Thompson and McHugh 2002). The purpose of the discussion here has been primarily oriented towards examining management as a form of domination and exploitation that has its origins within the rise and development of capitalism. The manner in which this is done, the strategies adopted, have clearly varied and have adapted to changes in capitalism and to other societal changes.

In examining the case studies this aspect of the nature of management provides a key to understanding how the organizations studied developed managerial approaches which contradicted the espoused purposes and philosophies on which they were established.

Chapter 5
Management and Class Analysis

Class and Social Division

The concept of class has been a major basis for sociological analysis, both theoretical and empirical, from the discipline's foundations. Its centrality has, in recent years, been subject to reconsideration, particularly in terms of other key social processes that give rise to divisions within society such as gender, ethnicity and sexuality (Lee and Turner 1996, Maynard 1994, Weber 2001, Devine et al. 2005, Kivisto and Hartung 2006). Our primary concern here is with the development of management in relation to its dominance within forms of organizing, and so we shall here confine the discussion to issues of class. As we shall see later, ethnicity played a major part in the establishment of, and the events arising within, both the Bus Garage Project and the Charlton Training Centre organizations examined in this book, and gender was an explicit issue in relation to the latter. For the present, we shall examine the issue of the class position of managers.

We saw in Chapter 3 that research within radical organization theory would need to take account of the class position of managers. We have seen how the occupation of management has grown over the past century, in terms of the proportion of the workforce categorized as 'managers' in various censuses and surveys. How might we understand this in terms of class analysis?

Marx's analysis of the class system under capitalism is well-known, that is, that there were and would increasingly be only two main classes, the capitalist class or bourgeoisie, and the workers, the proletariat:

> The history of all hitherto existing society is the history of class struggles. … Our epoch, the epoch of the bourgeoisie, possesses, however, this distinctive feature: it has simplified the class antagonisms. Society as a whole is more and more splitting up into two great hostile camps, into two great classes directly facing each other: Bourgeoisie and Proletariat. (Marx and Engels 1953: 33-34)

Such a direct and basic assertion about the class structure of society under capitalism has caused later followers of Marx problems regarding the position of the 'middle class(es)'. In particular the major growth in the proportion of the population which consists of waged/salaried non-manual workers, including managers, has been seen as an issue requiring analysis within a Marxist framework. As Wright put it, whilst all Marxists agree 'that manual workers directly engaged in the production of physical commodities for private capital fall into the working class … There is no such agreement about any other category of wage-earners' (Wright 1983: 30).

Marx's Analysis: Proletarianization

As discussed in the previous chapter, for Marx, the relationship between the labourer and the capitalist is exploitative, despite the fact that there is an apparently equal relationship between 'seller' and 'buyer' of labour power. Behind this appearance the capitalist mode of production enables the capitalist to realize surplus value, that is value in excess of that paid for the labour power of the worker. First, the labourer works under the control of the capitalist; secondly the capitalist, not the labourer, owns the labour power of the labourer, because it is a commodity which he has purchased. The purpose of the capitalist is to create surplus value, that is, to create value in excess of the use-value created by the elements of the labour process. To do this value in excess of the exact equivalent of the labour power used must be created.

Marx's analysis of the class nature of society under capitalism is thus based on this analysis of the relationship between the exploiter/capitalist and the exploited/labourer. The class structure of society is not static, but has developed from previous class structures of society, i.e. from feudal society. The bourgeoisie, exploiter/capitalist class, has revolutionized society replacing the previous forms of class domination with the 'cash nexus': 'for exploitation, veiled by religious and political illusions, it has substituted naked, shameless, direct, brutal exploitation' (Marx and Engels 1953: 35).

Classes left over from the pre-capitalist era are caught up within the revolution of social relations which the rise of capitalism brings. Thus the lower strata of the middle classes, lacking the capital to engage in modern industry on sufficient scale to compete effectively and because their specialized skills are rendered obsolete by new methods of production, sink gradually into the proletariat. So, say Marx and Engels, the proletariat is recruited from all classes in society, and society polarizes into the two great classes.

However the development of capitalism during the twentieth century does not seem to bear out a simple reading of Marx. That is, the proletariat does not appear to have absorbed all other classes except the bourgeoisie, nor to have become increasingly homogenous and unified. Instead, the advance of capitalism has brought into being new strata of middle class employees, educated, working apparently with varying degrees of autonomy and in many cases directing and controlling the work of others. This clearly poses a serious challenge to the Marxist analysis.

The New Petty-Bourgeoisie

One of the most significant attempts to deal with the 'problem' of the middle class(es) within a Marxist framework, was that developed by Poulantzas. He attempts to create a general framework for class analysis based on three fundamental premises. First, classes cannot be defined outside of class struggle:

> For Marxism, social classes involve in one and the same process both class contradictions and class struggle; social classes do not firstly exist as such, and only then enter into a class struggle. Social classes coincide with class practices i.e. the class struggle, and are only defined in their mutual opposition. (Poulantzas 1975: 14)

Secondly, class position is not dependent on the will of social agents but by objective position in the social division of labour: 'a social class is defined by its place in the ensemble of social practices, i.e. by its place in the social division of labour as a whole' (Poulantzas 1975: 14). Poulantzas insists that it is the analysis of the reproduction of these objective positions within the social division of labour (the 'structural determination of class') which is more important that the analysis of who occupies given positions. Thirdly, Poulantzas argues that this structural determination of classes takes place not only at the economic level but also at the political and ideological levels: 'the places of political and ideological domination and subordination, are themselves part of the structural determination of class' (Poulantzas 1975: 16).

Classes are then structurally determined by relations of production/exploitation (the economic), by relations of political domination/subordination (the political), and by relations of ideological domination/subordination (the ideological). Poulantzas then outlines a theoretical strategy for analysing class boundaries using and expanding on the three-fold set of criteria. He arrives at the conclusion that under modern capitalism we see four classes: the bourgeoisie, the proletariat, the old petty-bourgeoisie and the *new petty-bourgeoisie*. The new petty-bourgeoisie consists of white-collar workers, supervisors, technicians and the like, and is, Poulantzas argues, unified with the old petty-bourgeoisie.

He argues this in two stages. First, he discusses the economic, political and ideological criteria in relation to the new petty-bourgeoisie and the proletariat. Then he goes on to discuss these criteria with regard to the old petty-bourgeoisie and the new petty-bourgeoisie. At the economic level he uses the distinction between productive and unproductive labour to define the boundary between the proletariat and the new petty-bourgeoisie. Productive labour is, for Poulantzas, 'labour that is directly involved in material production by producing use-values that increase material wealth' (Poulantzas 1975: 216).

Wage labour is not in itself sufficient to determine class location. Since much of the new petty-bourgeoisie is not involved in such production of use-values it cannot be identified with the proletariat. The new petty-bourgeoisie *is* exploited by capital but through unproductive labour. At the political level the new petty-bourgeoisie is involved in supervising and controlling labour in order to ensure the continued extraction of surplus-value on behalf of capital. Moreover the new petty-bourgeoisie is also dominated by capital and so is excluded from both the proletariat and the bourgeoisie. At the ideological level, Poulantzas employs the distinction between mental and manual labour. In order for the reproduction of capitalist social relations of production knowledge of the productive process must be monopolized by capital and 'hidden' from the workers.

What is important for Poulantzas is not the actual content of the knowledge but its cultural significance in legitimizing the exclusion of productive workers from the planning and direction of the productive process. This 'cultural symbolism' extends from the traditional esteem placed on clerical work in general to modes of speech deemed 'proper' including the distinction between general culture and specific technical skills. However the knowledge is fragmented and dominated by capital, so the new petty-bourgeoisie is ideologically subordinate to the bourgoisie. The unity between the old petty-bourgeoisie and the new petty-bourgeoisie exists, according to Poulantzas, because of the common political and ideological positions. This arises because they have a common place in the class struggle:

> If the traditional and the new petty-bourgeoisies can be considered as belonging to the same class, this is because social classes are only determined in the class struggle, and because these groupings are precisely both polarized in relation to the bourgeoisie and the proletariat (Poulantzas 1975: 294).

He then goes on to describe the aspects of this common ideological/political position, in respect of reformism, of mythologies regarding social promotion and 'neutrality' of the state, and of individualism.

Wright's Critique of the Poulantzas Thesis

Poulantzas's analysis has been criticized most notably by Erik Olin Wright. He first provides a critique of the logic of the analysis of the boundary between the proletariat and the new petty-bourgeoisie. For Poulantzas this boundary is based on the distinction between productive and unproductive labour, productive labour being restricted to labour which both produces surplus value and is directly involved in the production of physical commodities. Wright regards this as an arbitrary assumption:

> If use-values take the form of services, and if those services are produced for the market, then there is no reason why surplus-value cannot be generated in non-material production as well as the production of physical commodities. (Wright 1983: 46)

Moreover, argues Wright, the distinction cannot be used as a criterion for class determination because actual positions contain a mix of productive and unproductive labour. He cites the example of the manufacture of packaging, which is productive labour in that, along with the contents, the packaging is a commodity, but is unproductive labour in that packaging forms part of the advertising for the commodity.

Most significantly, for Wright, the criterion Poulantzas uses cannot be used to separate the 'middle class' from the proletariat because the implication that they have fundamentally different class interests at the economic level does not hold true. The implication is that the new petty-bourgeoisie have no objective interest in socialism. It is true that there are divisions of immediate economic interest between various groups of workers, but these are not sufficient to imply an interest in perpetuating the exploitative capitalist system: '[n]one of these divisions changes the fundamental fact that all workers, by virtue of their position within the social relations of production, have a basic interest in socialism' (Wright 1983: 49).

Wright goes on to examine the way Poulantzas uses the political and ideological criteria in his analysis of class determination. Although Poulantzas insists that economic criteria play the principal role in class determination, Wright argues that Poulantzas's analysis does not bear this out. By schematically presenting the various criteria applied to the various classes in the analysis, Wright shows that the proletariat and the bourgeoisie can be seen to be polar opposites, that is, on each criterion they have opposite signs. In Poulantzas's analysis any deviation from the criteria defining the proletariat is sufficient to mean exclusion from the proletariat. So, Wright argues, under Poulantzas's analysis, 'an agent who was like a worker on the economic and political criteria, but deviated on the ideological criteria, would on this basis alone be excluded from the proletariat' (Wright 1983: 51). Therefore, argues Wright, the political and ideological criteria are effectively equal in Poulantzas's analysis, because they are always capable of pre-empting the determination of class at the economic level.

Wright also questions Poulantzas's use of political and ideological criteria. The core political criterion is position within the supervisory hierarchy, supervision being conceived as 'the direct reproduction, within the process of production itself, of the political relations between the capitalist class and the working class' (Poulantzas 1975: 227-228). But, says Wright, supervision can be conceived alternatively, as

> one aspect of the structural dissociation between economic ownership and possession at the economic level itself. That is, possession, as an aspect of ownership of the means of production, involves (to use Poulantzas's own formulation) control over the labour process. (Wright 1983: 52)

Not only has possession become dissociated from economic ownership under monopoly capitalism; equally possession has itself become internally differentiated. Control over the entire labour process (top management) has become separated from the immediate control of labour activity (supervision). Thus, argues Wright, supervision should be seen as a differentiated element of economic relations rather than a reflection of political relations within the social division of labour.

Wright questions Poulantzas's use of the mental/manual division as a determinant of class boundary. There are many *ideological* dimensions to divisions within the working class, including sexism, racism, and nationalism. Thus for example, sexism is a dimension of ideological domination/subordination within the social division of labour. This puts men in the position of ideological domination over women but does not thereby make a male worker *not* a worker. It is not clear why the mental/manual division should be chosen as the 'essential axis of ideological domination/subordination'. Wright rejects Poulantzas's argument about the class unity of the new petty-bourgeoisie and the old petty-bourgeoisie. On both the economic and the political levels they are in opposing positions. The old petty-bourgeoisie is constantly threatened by the growth of monopoly capitalism, whereas the new petty-bourgeoisie depends on monopoly capital (economic level). The old petty-bourgeoisie is opposed to the expansion of the state and to large state budgets, whereas the new petty-bourgeoisie has an interest in these.

Wright also criticizes Poulantzas's attempt to identify the old and new petty-bourgeoisies at the ideological level. Far from being unified the ideologies can be characterized as opposed. Individualism for the old petty-bourgeoisie is different from that of the new petty-bourgeoisie. The former individualism is concerned with individual autonomy, 'being your own boss', and so on; the individualism of the latter is careerist individualism, concerned with bureaucratic advancement. However the main objection to the thesis of unity between the old and new petty-bourgeoisies is that even if the ideologies were identical this would not be sufficient to call them a single class.

Wright: Contradictory Class Locations

Wright offers an alternative conceptualization of class boundaries, which he considers deals with the problem of ambiguous positions. He suggests that such ambiguities can be dealt with regarding such positions as occupying 'objectively contradictory locations within class relations'. Such contradictory locations should be examined in their own right; the contradictions cannot be eradicated by artificially classifying positions within the social division of labour unambiguously into one class or another. In fact Wright is concerned that class analysis, and the elaboration of criteria for such analysis, should not consist of the construction of formal, abstract typologies and the 'pigeon-holing' of people within such a typology. 'To fully grasp the nature of the class structure of capitalist societies', he argues

> we need first to understand the various processes which constitute class relations, analyse their historical transformation in the course of capitalist development, and then examine the ways in which the differentiation of these various processes has generated a number of contradictory locations within the class structures of advanced capitalist societies. (Wright 1983: 62)

Wright identifies three interconnected structural changes in the course of the development of capitalism: the progressive loss of control, by direct producers, over the labour process; the elaboration of complex authority hierarchies; and the differentiation of various functions originally embodied in the entrepreneurial capitalist.

The first change is a well established aspect of capitalist development, and indicates the first process underlying the fundamental capital/labour relationship. Even apparent deviations from the such loss of control by the worker over the labour process, for example by 'job enrichment' programmes or during the development of new technologies, cannot be sustained as objections. Experiments in worker participation and the granting of enlarged autonomy have always been undertaken within narrow limits with the objective of increasing productivity. The advances in, for example, the computer technologies have seen the deskilling of jobs so that operators who previously tended to be engineers now require little post-school education or training. Since capital attempts to extract as much actual labour from the worker as possible capital must try to gain control over the labour process. The fact that under different conditions this may be accomplished by varying degrees of direct control does not alter this.

In discussing the process by which the functions originally embodied in the entrepreneurial capitalist are, within the development of capitalism, differentiated, Wright counters the view that there has been a separation of ownership and control. Such a view is central to the 'managerial revolution' thesis which argues that modern corporations are no longer controlled by capital but by management within organizational hierarchies. But, says Wright, hidden behind the apparent separation of ownership and control lies a complex process which involves a whole series of structural transformations and differentiations. He identifies three dimensions of ownership: formal *legal* ownership, *economic* ownership ('control over the flow of investments into production') and *possession* ('control over the production process').

There has been a partial separation between economic ownership and possession, arising from the concentration and centralization. This is caused by the competitive pressures on capital, making it less practical for ownership and possession to be undertaken within the same function, and also because the tendency within monopoly capitalism for the centralization of economic ownership to proceed more rapidly than the concentration of possession. There has also been a gradual dissociation between formal legal ownership and real economic ownership, through the dispersion of share ownership. However, rather than drawing the conclusion that the supporters of the 'managerial revolution' thesis draw, Wright concurs with De Vroey (1975), who argued that the dispersion of stock was a means for *reinforcing control* by big stockholders. This enabled them to command an amount of funds that was out of proportion to their actual ownership.

The process of concentration and centralization of capital also gives rise to various forms of differentiation within the dimensions of economic ownership and possession. Relations of possession are concerned with the direction and control of the production process under capitalism and can be analysed into two distinct aspects, the control of the physical means of production and the control of

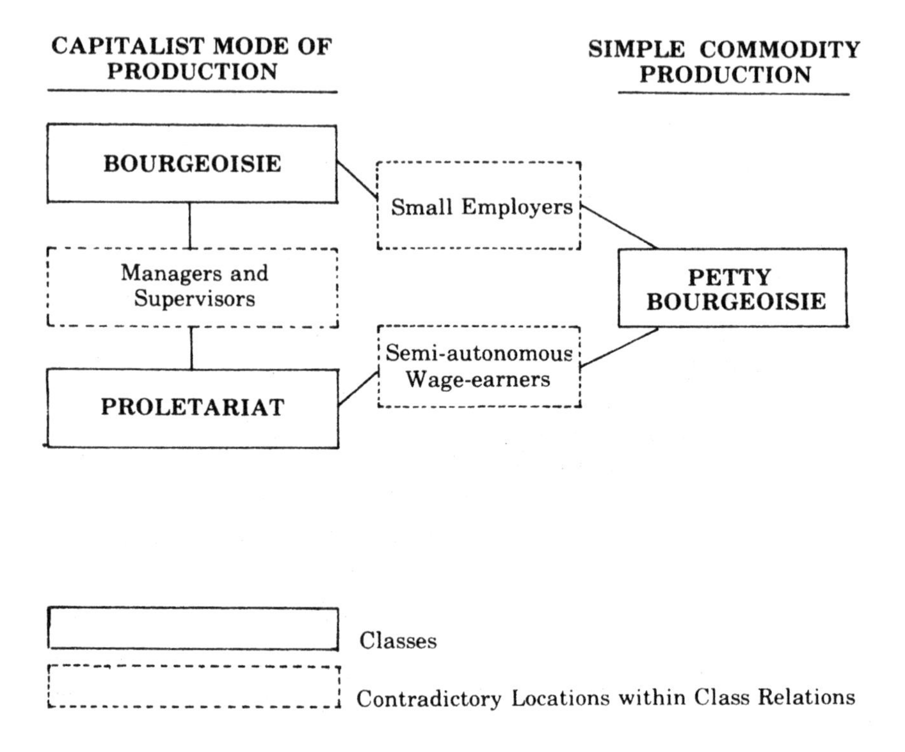

Figure 5.1 Wright's model of class locations

Source: Figure reproduced from Wright (1983), courtesy of Verso.

labour. As capitalist enterprises expanded, layers of supervision were introduced so that complex hierarchy of control resulted. Such hierarchy of control exists over both aspects of possession. Similarly in respect of relations of economic ownership the degree of participation in the control of financial capital is organized hierarchically, with the highest executives usually also having substantial degrees of legal ownership. Wright thus isolates from this historical development three central processes underlying the basic capital–labour relationship, control over the physical means of production, over labour power and over investments and resource allocation. Wright stresses that 'these three processes are the real stuff of class relations in capitalist society; they are not merely analytic dimensions derived from a priori reasoning' (Wright 1983: 73).

From this analysis of the central processes Wright develops an analysis of contradictory locations within the class locations. Whereas the bourgeoisie, petty bourgeoisie and proletariat have clear class locations, managers and supervisors, small employers and semi-autonomous wage-earners occupy contradictory class locations within class relations (see Figure 5.1). For our purposes here, we shall concentrate here on the contradictory locations between the bourgeoisie and the proletariat. Such positions have a variable rather than an all-or-nothing quality, some being located around the boundary with the bourgeoisie, others around the boundary with the proletariat. In such locations are top managers, middle managers, technocrats, and line supervisors/foremen.

Top managers have full control over both aspects of possession (over the physical means of production *and* over the labour power of others) and partial control over investments (economic ownership). Middle managers have partial control over control over the physical and human means of production and minimal control over investment. Technocrats are the various technical and professional workers who have some autonomy over their own work, possibly some control over the work of subordinates but no control over the physical means of production. Closest in location to the proletariat are foremen and line supervisors. The only criterion for separating them from the proletariat is the ability to exercise control over the work of others. However even this has been reduced under the development of capitalism with the institutionalization of power, that is the establishment of rules and procedures to replace personal, arbitrary command.

So far Wright's analysis has focused on class positions directly determined by relations of production. He also examines briefly the class location of those who are not employed, in particular students, pensioners, and those who are 'permanently' unemployed existing on welfare benefits. However, most significant for our concern here, is the analysis of those who are employed but in political and ideological apparatuses, the 'superstructure'. These include various local and national state institutions and agencies, with the police force and education as the two examples quoted by Wright. In order to identify the class location of such workers, Wright invokes the notions of 'class interests'. His concern is to ascertain the relationship of these positions to the structural categories which have been defined directly by the social relations of production. Class interests are those objectives of class actors which are potential, and under certain conditions, actual objectives of class struggle. A distinction can be made between 'immediate' and 'fundamental' class interests. The former are those that arise within any given structure of social relations, taking the mode of production as give. Fundamental interests centre on those that question the structure of social relations, the mode of production.

Immediate interests are 'incomplete' interests. Struggles based on immediate interests, for example, over wages, reflect a correct understanding by workers of their immediate conditions of existence under capitalism. However, to restrict all struggle to such immediate interests would reflect an incomplete understanding of the nature of capitalist society as a whole. At the level of immediate interests there is considerable division among the working class, because of differing positions of advantage/disadvantage in the labour market: skilled versus unskilled workers, male versus female, white versus black and so on. So 'the durability of capitalism depends, in part, on the extent to which struggles over fundamental interests are displaced into struggles over immediate interests' (Wright 1983: 91).

Coincidence between immediate and fundamental interests of the working class occurs only in revolutionary situations, and indeed, argues Wright, this may be part of the definition of a revolutionary situation. Wright next expands the analysis from economic class interests (capitalist versus socialist modes of production) to political and ideological class interests (capitalist versus socialist organization of state and ideology). At the political and ideological level, the fundamental interest of the capitalist class, he argues, is to prevent the working class from gaining state power and achieving ideological hegemony. On the other hand, 'the fundamental interests of the working class at the political and ideological level are, in dialectical manner, to obtain state power and establish ideological hegemony' (Wright 1983: 95).

The fundamental interests of the capitalist class implies, argues Wright, the maintenance of hierarchical and bureaucratic structures within the political and ideological structures. By contrast the fundamental interests of the working class implies a qualitative restructuring of the state, and although the exact nature of such reorganization cannot be specified in advance, 'the minimum requirement is that they be *radically democratic and antibureaucratic*' (Wright 1983: 95, emphasis added).

From this analysis of antagonistic class interests Wright identifies three functional categories of positions within the political and ideological apparatuses:

- *bourgeois positions* which involve control over policies in the political apparatuses and the production of ideology in the ideological apparatuses;
- *contradictory locations* which involve the execution of state policies and the dissemination of ideology;
- *proletarian positions* which involve complete exclusion from either the creation or execution of state policies and ideology.

Wright's analysis thus provides a systematic approach for theorising the nature of 'intermediate' class locations under advanced capitalism, which Wright has used for empirical research on the class structure of the USA.

His analysis has been criticized by some writers. Ehrenreich and Ehrenreich (1979) argue that Wright is being too simplistic in regarding the working class has having 'determinate' class position thereby positing that the working class is inherently socialist. However this criticism is based on an earlier paper by Wright and is made in the context of their defence of their conclusion that there is a new Professional–Managerial Class, which is different from the 'new middle class' (white-collar workers such as sales and clerical staff) and from the older petty-bourgeoisie. Abercrombie and Urry (1983) point out that there are difficulties in using isomorphic categories as a basis for an approach which is gradational.

Wright acknowledged many of the criticisms. He did himself revise his own analysis (Wright 1985), whilst maintaining his claim that it was an advance over other attempts at addressing the 'problem' of the middle class. The concept of contradictoriness proved itself to be problematic, he argues, in the case of semi-autonomous workers and small employers. It was difficult to answer satisfactorily, he says, the question of why semi-autonomous workers should be viewed as having internally inconsistent interests. Similarly, whilst small employers 'may face various dilemmas in competing successfully in a world of large corporations ... this does not obviously imply that they have internally contradictory interests' (Wright 1985: 53). He also criticizes his original analysis in respect of the use of the concept of autonomy as a class criterion, but this does not apply to the position of managers. However, Wright explicitly excludes managers from the first criticism and his second does not apply to the placing of them in a contradictory class location.

Wright goes on to raise the difficulty of applying his original framework to classes in post-capitalist societies (Wright 1985: 55), and argues that he had shifted the emphasis in his framework from a focus on exploitation, central to Marxist analysis, to that of domination. These issues are more related to his attempt to develop a *comprehensive* approach to class analysis, capable of being applied effectively to empirical investigation in existing societies. He develops a more extensive framework with 12 class positions. Yet within this new framework, Wright maintains that 'these "new" middle classes do constitute contradictory locations, or more precisely, contradictory locations within exploitation relations' (Wright 1985: 87). He continues: '[w]ithin capitalism, the central contradictory location within exploitation relations is constituted by managers and state bureaucrats' (Wright 1985: 89).

Management and Contradictory Class Locations

Wright's analysis refutes the assumption implicit in 'orthodox' management thought, that management is a unitary phenomenon. That is, orthodox management theory, and management education and training, assumes that 'management' is essentially the same whether it is management undertaken by top managers in monopoly commercial enterprise, by lower level managers and supervisors in such enterprise, by top policy makers in institutions in the 'superstructure', or by lower levels of administrators and managers in such institutions. There may be differences of emphasis or of degree, but the overall nature of management is assumed to be the same. This can be seen perhaps most clearly in those management training approaches which are provided on an organization-wide basis, and in the hierarchy of educational programmes and qualifications, ranging from supervisory management programmes, through first degree and postgraduate to the MBA (and increasingly the degree of Doctor of Business Administration).

It may also be seen in the way that the non-productive public sector, such as education, public housing, health and social services, have increasingly been required to develop management approaches. The assumption that 'management' is the same wherever it is undertaken serves to mask the fundamental class interests of different social groups, either by inducing 'false' perceptions (i.e. that the class interests of managers are essentially coincident with that of capitalists) or by presenting management as in some way separate from issues of class interests and class struggle. Wright's analysis emphasizes the need to start from a recognition of the conflict in fundamental class interests arising from the relations of production within the capitalist mode of production. The functions of supervisors and first-line managers, having no control over investments or over the physical means of production, and only minimal control over the labour power of others, are different from middle and top managers. Their fundamental class interests lie most closely with the working class, although this is masked by the conflicts of immediate interests. The question of the class interests of middle managers is much more problematical. Wright remarks that the contradictory quality of their class location 'is much more intense than in the other cases we have discussed, and as a result it is much more difficult to assess the general stance they will take within class struggle' (Wright 1983: 79).

However, capitalism is itself essentially riven with contradictions, and is subject to periodic crises of overproduction and collapses in the capacity to accumulate surplus value. Under these conditions the privileges that middle managers obtain from capitalism come under increasing pressure, with redundancy a real prospect for many as firms are rationalized, taken over or collapse. Moreover the counter-pressure from labour against capital's attempts to maintain the level of surplus value extorted, is most keenly felt by middle managers. Under such conditions the prospects for radical change in the form of the replacement of capitalist mode of production for socialism become heightened, and middle managers may be more open to reconsideration of where their fundamental class interests lie, and in that critical moment decide that they lie with the socialist camp. Whether or not they do so decide, or opt for alliance with capital, or merely become 'dissolved' in the process of revolution, will depend to a significant degree on the success with which an alternative approach to organization and coordination of production, the 'technical' as opposed to the 'social control'/'ideological' aspects of management, is articulated. Moreover, Wright's analysis of the class location of intermediate strata in political and ideological structures effectively challenges the image of management in these structures as a neutral-technical function, serving the needs and wishes of society expressed through the processes of the democratic system of party politics.

Events in the public sector have indicated how the workers in lower levels of the contradictory class locations have engaged in struggles based around their fundamental class interests identified as coinciding with those of the working class. Thus, at various times, cuts in education, health services, social security have led to campaigns not only to resist such cuts, but also to change the very nature of such services so that they meet the real needs of those for whom they are supposed to be provided. These campaigns have, of course, had limited success partly because of the way they have been conducted within their separate spheres, and partly because of conflicts in immediate interests. Thus the industrial action by teachers, or by health workers, for example, often gain considerable public sympathy coupled with public animosity, because of the disruption to the services.

Of critical importance for middle managers in such institutions is the very nature of those institutions. Unlike productive enterprise, institutions in the superstructure do not have profit and the competitive nature of the market as sources of legitimation. There is therefore likely to be much more conflict over the purpose and nature of such institutions, especially given the bureaucratic nature of relations of employment: significantly more job security for employees and highly detailed procedures for dealing with recruitment, discipline, grievance, and remuneration. Moreover, such positions are usually associated with values of professionalism and public service which may come into conflict with pressures on such middle management workers to limit provision of services and to adopt management practices associated with private enterprise. Thus, under conditions of intense crisis there may be a real chance that such workers will ally themselves with the working class. However, again, whether or not they do so will be significantly affected by the degree to which the alternative is articulated.

Of course, the prospects for radical change in the mode of production with its particular relations of production, and for radical change in the nature of political and ideological structures, are integrally connected. Such change would constitute a change in the nature of class rule, in the nature of the state itself. We will therefore next examine the nature of the state, and its implication for a radical perspective on management.

Chapter 6
The Capitalist State and Management

Theories of the State

In his review of the literature on managerial effectiveness, undertaken for the Manpower Services Commission (MSC), Hales (1980) points to a particular problem for government involvement in management training. Government must seek to avoid being seen as a neutral arbiter between differing sectional interests, but by promoting management training it may be promoting the interests of one particular sectional interest.

> In other words, for government 'neutrally' to promote management training presupposes that it is accepted that: (i) The 'technical' view of management correctly prevails; and (ii) That both the fact and the implications of the separation of ownership and control have been proved. (Hales 1980: 24)

As we have seen such presuppositions are challenged by radical views on management. Moreover, the role of the state as a 'neutral referee' arbitrating between sectional interests is itself a major issue which a radical view would challenge. Such a conception of the state (or more usually, 'government') is referred to as the 'pluralist' view. This view argues that, while it would be simplistic to regard every citizen as having equal political power, nevertheless within Western political democracies the outcome of political processes is one which more-or-less reflects the wishes of many conflicting groups (for example, Dahl 1961, Polsby 1963). Insofar as conventional management literature acknowledges the existence of political processes it is such a view which tends to be assumed. Much management textbook literature merely views the state and/or government as part of the 'environment' of the organization, the 'political' aspect using the PEST model commonly used in such textbook literature for analysis of an 'organization's environment' (that is, political, economic, social and technological 'factors').

A textbook on management in central and local government explains that

> the study of management is the systematic examination of the ways in which groups of people organise themselves to achieve goals by co-operative action. The purpose of this examination is to try to develop ideas and suggestions for improving the efficiency and effectiveness of such co-operative activity, whether it be running of a government department, a hospital, a business firm or indeed any other organised group'. (Bourn 1979: 1)

Thus the pluralist view of the state and the rational-technical view of management fit well together. Government as 'neutral referee' arbitrates between competing interests in society, 'professional' managers obey the letter and spirit of laws passed and have 'regard' for the interests of various groups besides the owners of the enterprise, and management is a general activity concerned with improving effectiveness and efficiency of state organizations as well as private enterprise.

Such a view is often contrasted with 'elite theory' which argues that the major decisions in all spheres of society, business, the military, and the government, are made by a 'power elite' (for example, Mills 1956). The elite is shown to be made up of members of the upper economic class, who are linked in a variety of ways, including family relationship, membership of clubs and societies, common patterns of education (public school and elite universities), and linking directorships of significant organization, in particular finance institutions, large companies and quasi-governmental bodies. By itself, the mere existence of a ruling elite would not necessarily undermine the conventional, rational-technical view of management, because it could always be argued that such rule was functional for society, so that all members benefited and would be worse off under any other form of organizing state rule, or that such rule was an inevitable consequence of industrialization, as argued by Burnham (1941).

Marxist Theories of the State: Poulantzas

Within Marxist literature the theory of the state has been relatively neglected until recently. Marx himself did not develop a thorough analysis of the state, his principal theoretical purpose being to examine the capitalist mode of production. Poulantzas (1969) argues that later followers of Marx, principally in the Second and Third Internationals, held to a conception of the state as merely an epiphenomenon reducible to the economic 'base', a view he refers to as 'economism'. As a result non-Marxist, bourgeois analyses had gained prominence. He pays tribute to Miliband (1969) who deployed considerable empirical evidence to attack the arguments of the pluralists. Miliband concluded that there exists a plurality of elites which constitute a ruling class, which furthers its sectional interests through the actions of the state.

However Poulantzas considers Miliband's to be faulty in that he accepts the bourgeois, pluralist definition of the problem and that he fails to analyse the structural components of the capitalist state. Thus Miliband examines the cohesiveness of the capitalist class and the relationships between various elite groups. Poulantzas argues that, no matter how divided the ruling class may be, a capitalist state is still a capitalist state. The relationship between the capitalist class and the state is an objective relation, not dependent on whether or not members of the class directly participate in the government and state apparatuses:

> This means that if the function of the state in a determinate social formation and the interests of the dominant class in this formation coincide, it is by reason of the system itself: the direct participation of members of the ruling class in the State apparatus is not the cause but the effect, and moreover a chance and contingent one, of this objective coincidence. (Poulantzas 1969: 245)

Moreover, argues Poulantzas, the objective interests of the capitalist class is best served when the ruling class is *not* the politically governing class. Poulantzas conceptualizes the state as forming 'an objective system of special "branches" whose relation presents a specific internal unity and obeys, to a large extent, its own logic' (Poulantzas 1969: 248).

Analysis of the state must consider the particular form of relations among its branches and the predominance of one or more branches over the others. However this analysis must be undertaken with respect to the state in its unity, in particular in respect of modifications and changes in the relations of production and in the class struggle. Moreover, attention must be given not only to the repressive apparatuses of the state, which have been the main focus of much classic Marxist theory, but also to the ideological apparatuses. Unlike the repressive apparatuses which have a rigorous internal unity, the ideological apparatuses possess greater autonomy. Thus for Poulantzas, Marxist analysis of the state must recognize the relative autonomy of the state from the capitalist class, rather than seeing it as merely the direct instrument of capitalist class rule. The various branches of the state form a unity, which although having a certain degree of independence from the economic 'base', nevertheless can only be properly understood in terms of the total unity with the relations of production and class struggle.

How the Ruling Class Rules: Therborn

Therborn (1982) pays tribute to both Miliband and Poulantzas, and seeks to develop a Marxist analysis which returns to Marx's central task in *Capital* which was 'not to identify those who have wealth and those who are poor, nor those who rule and those who are ruled, but … to lay bare "the economic law of motion of modern society"' (Therborn 1982: 227). For Therborn the focus of the approach he advocates is

> on the historical social contexts and modalities of power, and the first question is: What kind of society is it? Then: What are the effects of the state upon this society, upon its reproduction and change? (Therborn 1982: 227)

The primary aim of this should be the effects the state has on the production and reproduction of a given mode (or modes) of production (Therborn 1982: 241).

In order to be able to analyse the effects of the state on the production or reproduction of the mode of production Therborn proposes outline typologies of state interventions and of state structures. The typology of state structures, argues Therborn, should distinguish between the extent that the rule of a particular class can operate through any given state structure under examination. He stresses that it should not be assumed that a particular state has, at any specific point in time, a homogeneous class character. In setting out an outline of a typology of state interventions Therborn argues that two main dimensions should be used: the effects upon given relations of production, and the effects on the relations of political domination. Interventions can further, merely allow, go against or (at the limit) break the given relations of production. The effect upon relations of domination can be to increase them, merely to maintain them, or to go against or (at the limit) break them. He then goes on to propose various ways of refining the model he proposes, in particular refinements which distinguish between effects on different fractions of capital. Therborn presents this outline framework of state interventions and state structures to show that the ruling class does not make, as a compact unit, all the important decisions in society but exercises its rule by 'a set of objectively interrelated but not necessarily interpersonally unified mechanisms of reproduction, through which the given mode of exploitation is reproduced' (Therborn 1982: 244).

He therefore goes on to examine in broad outline those mechanisms of reproduction. Referring to Poulantzas (1973) he points out the two aspects of reproduction of society: the reproduction of the positions of the given social structure and the reproduction of individuals who can occupy them.

> But also, new generations of individuals – and the given individuals year in and year out – have to be trained to occupy the given positions, to be qualified or subjected to fulfil adequately the tasks provided by the social structure. (Therborn 1982: 244-245)

Therborn identifies three broad types of mechanisms of such reproduction: economic constraint, political mechanisms for administration and for repression, and ideological mechanisms. The first mentioned is also of primary importance. Economic constraint serves to exclude certain relations of production making them non-competitive, and so makes other relations of production favourable and determines the range of political options. And the given structure of economic positions is reproduced in a constant process 'by sanctions of bankruptcies, unemployment, poverty, and sometimes outright starvation' (Therborn 1982: 245).

The political mechanisms of reproduction have two subtypes, administration and repression. Administrative interventions include taxation, regulations, social security policies, manpower policies, counter-cyclical policies, and so on. These operate to ensure that the substructures of society have an overall compatibility. Other administrative mechanisms are directly concerned with the reproduction of the state apparatus itself, such as constitutional arrangements, procedures for handling of issues, even legal conceptions themselves. Repression can be carried out through a variety of means. The police, army, prisons are obvious but Therborn points out that mental hospitalization has been used, and still is used, in certain states. Ideological mechanisms of reproduction have often been assumed by sociologists to be the only, or at least the main ones. Therborn shows that this is a mistake, but does see such ideological mechanisms as important. However he argues that the primary role of such mechanisms is not to legitimize the prevailing system but rather in the differential provision of skills and knowledge, and a differential shaping of aspirations and self-confidence. The apparatuses are various and not necessarily congruent with each other: they include the family, the educational system, the mass-media, the religious organizations, on-job training, the workplace. Through these mechanisms of reproduction, says Therborn, the ruling class can keep state power without having to supply the administrative and political personnel. Even the politicians, including those representing working class voters, are shaped by these mechanisms especially the ideological mechanisms.

State Theory and Management

We saw in the previous chapter that class analysis provides some insights into the relationship between management and the capitalist class, particularly in respect of the 'contradictory class location' of managerial workers. However we should remember that, as Therborn shows, the rule of the ruling class is effected through the state, not through subjective intentions and actions, but through mechanisms of reproduction of society with its particular mode of exploitation. Under capitalism the dominant mode of exploitation exists within the mode of production, but the political and ideological apparatuses of the state cannot be reduced to relations of production. Examination of the nature of management within the analysis of the state must therefore move beyond the debate over the labour process, and take account of these political and ideological mechanisms. Of course management takes place within various institutions of the state, and so management in the public sector requires such analysis. Moreover there is a large number and variety of organizations within the voluntary sector, which by its very nature can only be properly understood in terms of its relationship to state institutions. However before turning to these we will briefly examine some aspects of management within private enterprise, in terms of the foregoing analysis of the state.

Private Enterprise

We can observe three clear ways in which the hierarchical nature of management in private enterprise and the state are linked in the processes of reproduction of the relations of production. First, it provides for the apparent separation between ownership and control of the means of production. The 'managerial society' thesis, postulated by various theorists in varying ways (Berle and Means 1932, Burnham 1941, Dahrendorf 1957), emphasizes the changes in share ownership in capitalist enterprise to argue that this constitutes a fundamental change in the control of such enterprise. Nichols (1969) notes two strands to this thesis, 'sectional managerialists' and 'non-section managerialists'. The former view sees management as acting in its own sectional interest rather than being profit maximizing. The latter view sees management as disinterested and neutral, acting in a socially responsible manner arbitrating between competing interests. Nichols argues out that 'non-sectional managerialism has a high ideological potential for businessmen' (Nichols 1969: 55).

While the managerialists focus mainly on top management, the ideological importance extends to other levels of management. The notion of management as neutral, social responsible and undertaking disinterested arbitration is carried through the hierarchical management structures of organizations. Hierarchy provides the material sets of practices in which such ideology in embedded. Various state apparatuses are involved in this. The education system provides various levels of qualifications which are used as part of the selection methods for the various levels within organizational hierarchy. Selection processes are, of course, also 'rejection' processes thereby providing a framework for shaping the aspirations of potential recruits to the various hierarchical levels (we should also note the processes by which individuals self-select by *not* applying for positions which they do not expect to obtain – because of the selection-rejection criteria applied – thus promoting the perception that selection is a rational, neutral, technical process). The increasing emphasis on academic and vocational qualifications may be seen, not as necessary prerequisites for managerial work, but as the increasing use of the education system for the 'differential shaping of aspirations and self-confidence and in a differential provision of skills and knowledge' (Therborn 1982: 246).

Secondly, organizational hierarchy not only acts as an ideological process but also provides for an apparent separation between the political mechanisms of state rule and the actions of management at the workplace. Miliband (1969) argues that in order to ensure the continuation of capitalism's dominance when reform fails to achieve its promise, the state exercises the option of repression. Among the actions needed '[t]he independence of trade unions must be whittled away, and trade union rights, notably the right to strike, must be further surrounded by new and more stringent inhibitions' (Miliband 1969: 243).

Two facets of industrial relations legislation are worthy of note here. The first is the use of the concept of 'unlawfulness' to combat secondary picketing, introduced in the UK by the Conservative Government in 1980 and not repealed by Labour Government after election victory in 1997 and subsequently. The introduction of this concept of 'unlawfulness', rather than 'illegality', enabled the government to appear uninvolved in legal action taken against trade unions. Rather than banning such industrial action, that is, making it illegal, the government provided a legal framework in which employers are encouraged to use the courts to ban it through injunctions. Refusal by trades unions to comply results in actions for contempt of court with the possibility of unlimited financial penalties, and ultimately sequestration of assets. Such action was taken in a number of cases during the 1980s in the newspaper and mining industries; in each case, calls for the Conservative Government to intervene have been answered by the claim that the 'rule of law' must be maintained.

The second facet is the increasing use of massive, coordinated policing of disputes and the application of vaguely defined police powers, such as arrests for 'obstruction' and for 'conspiracy', and 'warnings' to people not to proceed to a picket. On the pretext of maintaining 'public order' and 'protecting individual liberty' (to go to work whilst a strike is in progress) the police repress workers in their disputes with employers. Significantly, the two major industrial conflicts that took place soon after the enactment of industrial relations legislation in the early 1980s, the miners' strike and the printworkers' dispute with News International, were concerned to maintain employment at a time when capital has been shedding labour in response to the crisis of profitability.

Thirdly, organizational hierarchy provides for an apparent separation between the economic constraints which the state uses to ensure the reproduction of the capitalist mode of production. The policy of reducing wage levels through mass unemployment (justified as the result of workers 'pricing themselves out of jobs') is effected through the collective investment, marketing, financial and other decisions of management. Lower levels of management are caught up within a context of economic constraint which has its origins in government economic policy. Redundancies, short time working, rationalization of plants appear to be and are treated as aspects of corporate strategy, but can be traced to state intervention in support of capital.

The Public Sector

The public sector includes various organizations which are concerned with production and provision of services on a commercial basis. The nationalized industries had a chequered history in respect of the level of subsidy which government is prepared to provide, or of the level of profit expected. In determining such levels various criteria and methods of calculation had been used, for example, for fixing prices when there is one major customer also state owned (as with deep-mined coal and the Central Electricity Generating Board), or in assessing the social costs and benefits (as with British Rail). However, certainly after the crisis of 1976 when IMF set conditions for loans to the British Government, nationalized industries were more and more often required to operate as if they were in the private sector, that is, to generate profit. In order to do this a major component of management strategy has been to shed labour on a mass scale and to increase the productivity of the remaining workforce. The privatization of nationalized industries by the post-1979 Conservative Government was undertaken after such management strategy had been put into effect. Thus the profitability of nationalized industry was a necessary part of the continuation of the capitalist state (to limit public expenditure and possibly to generate revenues) but was presented as possible only when 'management are allowed to manage', as demonstrated by the amazing turnaround in profitability accomplished by labour-shedding. The overall cost to society in terms of lost production and expenditure on unemployment benefits and so on was excluded from the debate, for it would seriously challenge the nature of the state and the mode of production which it would sustain.

Moreover as large sections of industry and public services are directly within the control of the state it is able to ensure the continuing reproduction of positions and occupants of those positions. This may be seen as playing two important functions. First, through strategies affecting mobility of management (salary levels compared with private sector, expansion and retraction, and so on) the state controlled sectors can replicate the structures of private sectors. Secondly, as a major client for management education and training, the nature of such education and training can be affected in ways which the state requires. The reorganization of two important areas of state controlled public services during the 1970s, the National Health Service and local government, was a major state intervention which, among its other effects, provided for significant alteration to patterns of management mobility. The result, in terms of the internal structuring of particular district, area and regional authorities and of particular local authorities, has been to increase the importance of general, corporate management compared to specialized area of expertise whether professional or geographical. This enables policy formulated at the centre to be effected locally, since the centre is linked to the periphery through the hierarchy of managers who occupy their positions because they have successfully emerged from the qualification-subjection processes of selection (recognized education, training, career history, performance appraisal and so on). Thus, for example, the implementation of policies of competitive tendering and outsourcing, whereby work that was hitherto undertaken by staff directly employed by an authority was turned over to commercial organizations, is the result of plans and decisions by management within the authority.

The effect of state controlled sectors on management education and training comes about in a number of ways. Direct provision of such education and training through training centres affects the pattern of mobility among those engaged professionally in management education, as part of the qualification-subjection mechanisms involved. Differential provision of such education and training (who obtains what kind of training, for how long, where, and so on) reinforces the hierarchical aspects of the particular state agency or sector which reflects the hierarchical nature of non-state sectors and is reinforced by patterns of mobility. And the 'purchase' of management education and training based on managerially defined needs affects both the structure and practices of management education and training. Thus the large-scale funding of specific programmes of management education and training is determined by the policy directions set by the state and responded to by the various institutions involved in such education and training. The opportunities for developing alternative forms of organizing work within the public sector such as collective working, are severely restricted, and so the opportunities for developing education and training for collective working are restricted.

The Local State and Community Work

The experience of community workers during the 1970s led many to question the nature of the 'welfare state' and the 'local state'. A major stimulus for this was the various initiatives of the government to deal with urban poverty, which studies in the late 1960s showed to be still existing despite the development of the welfare state and the post-war economic 'boom'. The Community Development Project was established in 1968 (Loney 1983), under a Labour Government, which aimed 'to find, through experiments in social action, how to effect a lasting improvement in social situations which display many symptoms of individual, family and community malfunctioning' (quoted in Cockburn 1977: 123-124). As Cockburn points out, the approach illustrated by this was one of social pathology, with the problem being located in the individual, the family and the community. The Project was set up as a collaboration between central and local government, with a number of local projects being established with a five year lifespan as 'a neighbourhood-based experiment aimed at finding new ways of meeting the needs of people living in areas of high social deprivation' (quoted in Community Development Project 1977: 4).

Each project was to have an associated research team based at nearby universities and colleges, funded by central government, and a central Community Development Project (CDP) team was based at the Home Office. *Gilding the Ghetto* was a report written by six workers in local projects (CDPs) describing and analysing the experiment. The report states that there were three assumptions underlying the CDPs' brief:

> Firstly, that it was the 'deprived' themselves who were the cause of 'urban deprivation'. Secondly, the problem could best be solved by overcoming these people's apathy and promoting self-help. Thirdly, locally-based research into the problems would serve to bring about changes in local and central government policy. (Community Development Project 1977)

However, the Project as a whole was ended in 1976 with some local CDPs being closed earlier. Various reports submitted to central government, dealing with issues such as industrial change, youth unemployment, and housing policy had been received with disinterest or hostility. Local teams working with local tenants and action groups came into conflict with councillors and officers of the local authorities. Contacts between local CDP workers developed and a central CDP Information and Intelligence Unit was set up in 1973, answerable to the teams. The nature of the focus for CDP workers changed so that they began to 'proclaim that the problems of urban poverty with which they were confronted, were the consequence of fundamental inequalities in the economic and political system' (Community Development Project 1977).

Shortly after major public expenditure cuts announced following IMF intervention in 1976, the central intelligence unit published a highly critical report; six weeks later the Home Secretary ordered the closure of the unit. As the various local projects were wound up, many workers attempted to develop an analysis of the whole Project. *Gilding the Ghetto* located the various government moves in the context of major changes in the economy and growing pressure from the working class. Industrial reorganization and rationalization in pursuit of maintenance and increase in profitability had resulted in growing unemployment which was becoming a *permanent* feature of the economic scene. Moreover, because significant forms of industrial rationalization included relocation into new geographical areas, unemployment was increasing more rapidly in inner urban areas, from which firms were relocating. The resulting changes in population structure in such areas had serious repercussions for the welfare state:

> For as the economic base of these working-class areas collapsed and the skilled, the mobile and the young moved out, the traditional family and community networks which had previously provided support for local people were badly undermined. (Community Development Project 1977: 37)

The state was being increasingly faced with growing demands on the services of the welfare state, both in the form of claims for unemployment and social security benefits and in the requests made to social services for assistance.

Moreover various outbreaks of youth violence and hooliganism, and political protest particularly concerning the Vietnam War (and for civil rights in the six counties of Northern Ireland) posed problems of social control for the state. In this context, argues *Gilding the Ghetto*, the CDP can be seen as an attempt to maintain social control without recourse to overt repression, which in any case, as Northern Ireland had shown, was expensive and provided no guarantee of success.

> The obvious lesson [of the situation in Northern Ireland] is that quite apart from making the state's interest and costing far less, prevention is better than repression because it is more successful. ... The Home Office programmes represented to large extent an attempt to breathe new life into the crumbling institutions of the family and the community in order to mobilise cheap, informal social control mechanisms. (Community Development Project 1977: 48-49)

The traditional means of law enforcement were still available if the experiment in the 'softer' option failed, but their use

> would be at much greater cost and would represent a set-back for the governing idea that Britain can remain an orderly, self-disciplined society, free of violence, discrimination and crime without fundamental changes to the existing economic structure. (Community Development Project 1977: 48-49)

Gilding the Ghetto argues that the failure of the state to deal with poverty was seen to be a technical or administrative problem, capable of resolution once the right methods are found. The emphasis was on management techniques:

> Real solutions are seen to lie, not in the realm of politics, nor in the provision of extra resources, but in improving administrative practice with modern techniques, like programme budgeting, corporate management, computers and cost benefit analysis'. (Community Development Project 1977: 54)

Cockburn similarly argues that 'CDP was designed to fit into the post-Maud phase of management-happy local government' (Cockburn 1977: 124).

The Maud Committee (Maud 1967) had been established by the Government to examine management in local government, reporting in 1967. The Committee had concluded that there were serious deficiencies in the management of local authorities because of the existence of too many committees and departments, over-specialized, inward looking, insufficiently coordinated, and so failing to meet the general needs of the local people. It proposed that a management board be established, consisting of elected members, which would formulate overall policy objectives, take executive decisions, and coordinate and supervise the authority's work. A clear line should be drawn between the responsibilities of elected members and those of the officers employed by the authority. It was proposed that a new senior officer post should be created, having authority over the departmental chief officers and responsible for overall coordination and efficiency of the council's work. By this, it was hoped, each authority would become a co-ordinated and purposeful organization.

Local government reorganization, in London in 1963, and over the rest of England and Wales in 1972, provided the context for introducing such changes in management, as did reorganization of the National Health Service in 1973. Corporate management approaches were thus introduced at the initiative of central government, based on the corporate management methods in large businesses. The 'community approach' adopted in the CDPs and other localized initiatives of the state was, argues Cockburn, not an alternative to corporate management but complementary.

> Corporate management had concentrated on the internal management structure of councils. The central state's 'community package' was to make good its shortcomings – first by reviving, renewing, reproducing the relations of authority; second by concentrating on implementing policies; third by providing the sources of information about the working class needed by management. (Cockburn 1977: 131)

Gilding the Ghetto points out that as well as being experiments on the residents of the areas covered, they were also experiments on and with local authorities themselves, and most especially, they were experiments on behalf of the state, both central and local. The fact that the state took up suggestions from the CDPs in a highly selective manner, rejecting those calling for increased resources, reveals its main interest – to streamline management of the poor (Community Development Project 1977: 60).

One main outcome of the CDP and other initiatives was that local authorities had employed an increased number of community workers whose role, according to *Gilding the Ghetto* 'has been, and remains to help the local state to solve its problems of maintaining credibility in the eyes of the poorer section of the working class and to manage them and its services more efficiently' (Community Development Project 1977: 61). Thus, the writers conclude that the state has re-organized as capital re-organized, but had also taken 'a lesson in management techniques from industry which makes it better prepared to meet the consequences of capital's activities' (Community Development Project 1977: 62).

The Voluntary and Community Sector

The discussion above is important for understanding the relationship between the state and the nature of organization and management in what is generally termed the 'voluntary and community sector', which would include the two organizations examined in this book. The Community Development Projects, and other projects such as Educational Priority Areas (Central Advisory Council for Education 1967, Smith 1987), had been established as state initiatives but one purpose had been to promote the development of neighbourhood-based groups and organizations. The emphasis was on self-help and participation, as an antidote to the inflexible bureaucracy of local and national government agencies. Lees and Mayo (1984) identify various problematic concepts, such as 'participation', 'community', 'self-help' and 'self-reliance', which enabled varying meanings to be attached. These might be used to promote less-costly approaches to welfare provision, by getting individuals to 'stand own their own feet', and/or to initiate voluntary work as an alternative to state provision: 'getting more out of the community' as *Gilding the Ghetto* put it.

Alternatively, participation might be seen as a goal in its own right, enabling those who are not participating in the pluralist democracy to do so, articulating their demands and developing alternative forms of provision of services which are more responsive and flexible than those adopted by the bureaucratic state. Such variation in meaning may in part explain why voluntary and community action has been seen as worth promoting by all major political parties, at national and at local levels. The voluntary sector, was championed by the Conservative government when it took office in 1979. The philosophy of self-help was seen as completely in line with the philosophy of 'rolling back the frontiers of the state'. However by 1983 considerable changes were taking place.

In 1983, the urban regeneration budget (Urban Programme) for Scotland was cut by 30 per cent and cuts in the budget for England were made later. The story might appear to be a repeat performance of the history of the Community

Development Projects. However there had been important developments since the mid-1970s, which significantly affected the relationship between the state and the nature of organization and management in the voluntary and community sector. Unemployment in the late 1970s rose rapidly as capital reorganized in the aftermath of the 1973 oil crisis, and increased by over two million during the first four years of the Conservative government, 1979 to 1983. The Labour administration had faced considerable industrial unrest culminating in the 'winter of discontent' in 1978-1979 as workers in the public sector engaged in industrial action against the pay policy imposed.

The Conservative Party's clear victory in 1979 was achieved after a campaign which emphasized the post-war record levels of unemployment ('Labour isn't Working'), the high levels of inflation experienced under Labour and attributed by the Conservatives to high public expenditure, overmanning in industry and the public sector, and the 'power of the union bosses'. They claimed to be able to restore economic stability and prosperity and restore 'law and order' by following monetarist fiscal policy, restrictive industrial relations legislation, and increasing the powers of and resources for the police. Capitalism was going to be allowed to work as it should, managers would be allowed to manage, and the 'lawless elements' in society would be contained and restrained. Such a mixture continued to be the main thrust of Conservative policy throughout the following decade, but the effects on the working class were enormous. Unemployment became no longer a feature limited to particular areas of disadvantage, as had been the case in the 1960s when CDP was initiated. It became acute in those areas, especially among the young and in areas of high ethnic minority populations. In 1980 and 1981 there were massive displays of unrest as riots took place in inner city areas of London, Birmingham, Bristol, Liverpool, and Leeds.

The police were unable to bring the disturbances under control for several days and even then the danger of fresh outbreaks continued and further rioting did take place. Indeed the findings of both official inquiries (Scarman 1982) and other reports (Kettle and Hodges 1982, Wallace, Joshua and Booth 1981) held that the methods used by the police, both in dealing with crime and in dealing with disturbances, were to blame to varying degrees. The riot in St Pauls in Bristol, 1980, and the 1981 riot in Brixton were both sparked off by attempts by police to arrest young blacks in a somewhat insensitive manner. The causes of the rioting were held by most commentators and by the then Secretary of State for the Environment to be the high level of deprivation in the areas affected. The government responded in two complementary ways, by increasing the funds available for dealing with urban decline, and by enabling police forces to adopt methods and hardware for tackling disturbances with immediate and overwhelming effect. Thus, as Miliband (1969) argued, the state adopted the two complementary strategies of reform and repression.

The strategy of reform was however applied in significant ways. The funding came from three main sources, Department of Environment, Manpower Services Commission, and private sector. Private sector funds were controlled by consortia of commercial organizations, or committees made up of senior representatives of such organizations with token involvement of local community representatives. Often the initiatives were mainly concerned with property development and showpiece ventures, and also linked with MSC schemes which will be discussed below. The government played a key role in promoting and encouraging such initiatives. The message being promulgated, with considerable media coverage, was that private enterprise was the best way to tackle the problems (rather than public sector investment and democratic direction and control). What was less clearly promulgated to the public was that private sector initiative was good for business. The resulting uplift for the depressed building industry was most welcome especially in the large companies which specialized in municipal contracts.

The Department of the Environment increased its budget for the urban renewal programme. Urban Aid was the provision of government funds to local authorities on the basis of 3:1, that is, the Department gave three quarters of the cost of particular schemes, the other quarter being met from rates. The overall budget was set by the Department for each authority and each application required Department approval. The increase in the overall Department budget was produced by creating two new categories of areas, Inner City Partnership areas and Urban Programme areas. The former were those deemed to be in most need, the latter the 'next level' of need. Having set in motion an ideological strategy (promoting an image of being responsive to the deprivation in the inner city areas) the state also adopted strategies of economic constraint and of control by administrative means. Economic constraint was effected by continuing the cuts in rate support grant.

The effect of this was to limit the extent to which local authorities could use their own funds for actions which they themselves wished to take according to the wishes which the ruling group on the councils deemed the local electorate had indicated through the electoral process. This was reinforced by severe and increasing penalties imposed on authorities that exceeded the limits dictated by the government. The administrative mechanisms then ensured that such funding carried with it the means for ensuring that not only did the Urban Aid system not interfere with the requirements of capitalism, but that it actively promoted the interests of capital. The key element in this was the ruling that the budgets for the areas of high need, the Inner City Partnership Areas and the Urban Programme Areas, should be distributed in such a way that about three quarters was spent on capital expenditure. This covers such items as purchase of equipment and fixtures and fittings, but most importantly covers the rehabilitation and redevelopment of property, both land and buildings. Only one quarter was to be spent on 'revenue expenditure' the term used to refer to operating expenditure. This area covers in particular the wages of staff involved in projects.

The limitation of funding to cover staff wages led to three types of results. One result was that the wages offered for particular posts were significantly lower than equivalent posts in either private sector or public sector organizations, resulting in a limitation in the range of potential employees applying for posts. A second result was that some voluntary organizations applied local authority job evaluation systems so that wage rates were roughly equivalent to jobs in local authority. However this led to reduced staffing causing existing staff to be overworked as a matter of routine: work overload was *structured into* the organization. Thirdly and most critically in terms of effecting the state's strategy was that voluntary sector organizations were driven to look to and in most cases to accept another state scheme for funding, the Manpower Services Commission's Community Programme. The special measures introduced by the MSC to deal with the mounting pressure because of increasing unemployment were developed and operated at first through a Special Programmes Division and the Industrial Training Boards, the statutory, sector-based training agencies. However during 1982 the Government abolished two-thirds of the Industrial Training Boards, and developed the special programmes into a more cohesive fashion.

The Community Programme was developed to deal with adult unemployment; the Youth Training Scheme was developed for school leavers. An appearance of decentralization was created by the development of Area Manpower Boards to oversee these programmes. These boards had employer, trade union, educationalist and voluntary sector representation. However the real control lay with the MSC and through it the Government and the ruling class whose interests it served. This was effected by incorporating trade unions in a way which took advantage of a key weakness in trade union organization. Trades unions are organized nationally and by industrial/occupational sector. Local geographical organization by trade unions, into trades councils, is generally weak; trades councils were ignored in the development of the Area Manpower Boards. Moreover the conditions under which particular projects could be set up were very tightly laid down, and Area Manpower Boards members were largely reliant upon information supplied by MSC staff which serviced the AMBs.

The Community Programme was established in a way that can be seen to serve the ruling class interests. It provided a maximum average wage per worker of £60 per week, which was a reduction of £29 from the maximum allowed under its predecessor, the Community Enterprise Programme. Despite evidence supplied by a number of groups active in the field of employment initiatives that such a reduction provided *no net reduction* in government spending the reductions still took place. This was because most workers on £60 would be entitled to various statutory benefits which would bring them above that if they remained unemployed. Such an act is understandable in the context of the declared view of the government that the unemployed had 'priced themselves out of a job'. Many critics pointed out that the Chancellor of the Exchequer had publicly referred to the government's desire to see unemployed people finding useful work to do in the community, in his budget speech in 1982. The MSC in fact introduced on a limited basis closely controlled by its Head Office a scheme whereby unemployed people undertook 'voluntary' work whilst remaining eligible for unemployment and/or supplementary benefit.

The 'Voluntary Projects Programme' was seen by many critics as a pilot project towards a scheme whereby the unemployed were compelled to do such work for the community in exchange for benefit. The reduction in the wage rate for Community Programme certainly provides evidence in support of this view. The procedures for establishing a project under the Community Programme ensured that the project was not to any significant degree a threat to the interests of capital. First, a handbook containing a whole set of rules, what must be done and what must not be done (not merely advice), weeded out any project with a campaigning purpose. These were deemed 'political' and were explicitly banned from the Programme. Also caught in this ban were printing workshops, because these had been used by community groups to print campaign literature. Several 'Unemployed Workers' Centres' were restricted in their activities because of this ruling, and so many were set up under limited alternative funding, in particular from Greater London Council and Metropolitan County Council funding.

Secondly after submitting a project proposal, a technical officer would examine the proposal to determine whether the amount of work to be carried out was 'realistic': whether *in the view of the officer* enough work would exist to justify the number of workers asked for. This was coupled with another ruling, on training within a project. The Community Programme was a 'work scheme' not a training scheme. Projects were supposed to take on, as far as was possible, experienced and qualified staff. Any training carried out was to be a subsidiary activity and limited in time spent to a maximum of 20 per cent of total working time. Thus the emphasis upon the productive work side of a project was enforced.

Thirdly, the number of workers per supervisor was examined and MSC officers attempted to keep the number high; a norm of ten workers to one supervisor was applied. This limited the amount and extent of personal contact and on-job tuition possible and tended to lead to a directive style of management. Moreover the rate of pay for supervisors was much higher than for workers. The MSC provided £122 per week but often this was topped-up by the local authority through the application of local authority job evaluation systems. Thus the hierarchical structure intrinsic to the capitalist mode of production was structured into Community Programme projects through the administrative mechanisms of the MSC.

Once a project had been set up under Community Programme funding further administrative rules ensured that the project continued to support the needs of capital. Stringent application of the rule that all MSC funded employees be 'cleared' by the Employment Division, either through a Job Centre or Professional and Executive Recruitment, demanded a high degree of organized administrative expertise. ('Clearance' entailed checking by the relevant agency that the individual met the unemployment criteria.) Failure to meet this rule would lead to the MSC withholding funds for the particular employee for the period during which clearance had not been obtained. Also, the MSC required a manual accounting system which was often not compatible with the normal accounting system operated by a project. This meant considerable work being necessary to 'translate' between systems.

Furthermore, any changes desired by a project to the agreed set up entailed considerable form-filling, discussions with MSC officers, and waiting time pending approval, thereby involving key staff in work related to maintenance of the scheme rather than development of the project in ways which based on the social objectives originally set. At the end of the period of a project, which was a maximum of 52 weeks, a renewal could be obtained but only upon submission of a new application. Any employees who had completed 52 weeks employment on a Community Programme project would be ineligible for continued employment. Exceptions could be made by special waivers granted by the relevant MSC officer, but such waivers were seldom given. Thus a major upheaval was caused to a project because of the changeover of staff and the work needed to replace such staff: advertising, interviewing, induction, training, wages administration and so on. The net result of such administrative mechanisms was that Community Programme projects not only posed no threat to but also reinforced the relations of production implicit within the capitalist mode of production. These include hierarchical organization, division of conception and execution, and the predominance of productive work over the meeting of social objectives.

In this way organizations in the voluntary sector were incorporated into state mechanisms intended to meet and further the needs of capitalism and the interests of the ruling class. The appearance of independence on the part of voluntary and community organizations masked the reality that these projects were being managed on behalf of the state in order to maintain and reproduce capitalist relations of production. We shall examine the nature of the MSC and its schemes when we analyse the context in which the organizations in the field studies were established.

Conclusion

The foregoing examination shows then that there is considerable scope for placing an analysis of management within the context of state mechanisms. Such mechanisms are related to the mode of domination in a class divided society, which under capitalism is based in the relations of production emerging from the capitalist mode of production itself. This can be seen at work not only in self-proclaimed capitalist economic enterprises but also 'enterprises' which are intended, in theory at least to be concerned with objectives which are not based on the production of surplus value. Neglect of such analysis of management and management education and training can only lead to the continuation of the dominance of capitalism. The analysis may lead to the development of alternative approaches to participatory forms of management for a socialist mode of production, a key feature of which must be the alternative ideology and the education and training approaches compatible with such ideology.

Chapter 7
Popular Planning in Modern London: A Context of Disparities

Context and Totality

As discussed earlier, radical analysis of organization and management must be located within an analysis of the wider societal totality. As Burrell and Morgan state, 'totality' implies 'that organizations can only be understood in terms of their place within a total context, in terms of the wider social formation within which they exist and which they reflect' (Burrell and Morgan 1979: 368). The total context for the analysis of the two organizations studied here involves, in particular, the significant contradictions and changes in the relations of production, class relations, and the state; these must be identified and related to the tendencies and events in particular concrete situations. Thus an adequate understanding of the emergent events in both the Bus Garage Project and the Charlton Training Centre can only be achieved by uncovering the processes of determination arising from within the wider totality. Before examining the details of the events described we will therefore examine the context in which they were situated.

Changes in Capitalist Production and Effects on Employment

It is important to note that both organizations were in London, the capital city of the United Kingdom and, in popular understanding, the centre of affluence: the rich 'south' contrasting with the industrial decay and under-development of other urban and rural areas of the United Kingdom. However, although the overall rate of unemployment for the Greater London area is, and was during the period of the study, lower than the national average, that overall rate masks the fact that there are major areas of high unemployment within London that greatly exceed the national average. Moreover London, in 1985 (a key time period for the organizations studied in this book), had the highest urban concentration of unemployed people in the industrialized world, with a total of over 400,000 (Greater London Council 1985: 65). The total had risen to this figure from 329,000 in May 1981, when the Labour Party had gained control of the Greater London Council from the Conservative Party. In fact the later figure was calculated in a different and more restrictive way from the earlier, as the Government had changed the basis of calculating the number of unemployed several times during the intervening period (Levitas and Guy 1996). By contrast the number of unfilled job vacancies was fewer than

30,000 in 1985, and the figure had hovered between 20,000 and 30,000 during this period.

Although the increase in unemployment after 1981 was rapid, it was the continuation of a trend that had been taking place since the 1960s. The trend was of decreasing employment, and unemployment was not the result of increased population. In fact the population of London had been declining since the 1940s, and the populations of both Greenwich and Brent had decreased over the decade from 1974 to 1984 (Greater London Council Intelligence Unit 1986, based on Census data). Between 1961 and 1983, the number of people employed in London's factories fallen from 1.43 million to 594,000. The increase in unemployment had come about, then, because of decline in the demand for labour. The GLC concluded from analysis of the changes in the demand for labour over the previous two decades, that

> Within twenty years the size of London's job economy has fallen by a quarter, factory work has fallen to a tenth of overall employment, about a third of Londoners now work for the state, about 40 per cent work in offices, and nearly 20 per cent work in banking, finance and business services. (Greater London Council 1986: 46)

Five main factors were identified as being behind these changes. There had been a regional shift away from the centre of London, and the new factory sites built in the 1920s and 1930s in west, into the wider south east region. There had also been in the nature of employment away from production into the knowledge economy. Changes in the London economy's role in the international division of labour, particularly its 'Europeanization' which had led to a growth in intermediate business services but the loss of manufacturing to new centres in mainland Europe. There had also been changes in consumer services, both in private sector and in public services where there had been cuts in public sector employment because of government policy, intensified by privatization. The particular affects of the economic depression on London, which hastened trends, with employers using the traditional methods of restoring profitability: reducing wages, increasing intensity of labour, cutting public services, relocating to areas of cheaper, less organized labour.

Of course London was not the only city to suffer. Between 1960 and 1981 the major conurbations in the UK had lost 1.7 million manufacturing jobs, 70 per cent of the total job loss nationally of 2.1 million (Greater London Council 1985). The 'London Industrial Strategy' pointed out that depression does not take hold in an undifferentiated way, but takes hold of *particular* areas. In London de-industrialization had been concentrated in inner London and the east. Major employers of previous periods had been run down or had closed. Male unemployment in Greenwich was approaching 30 per cent by 1985. Moreover the rate of unemployment was much greater in some groups than others, in particular among the young, among black people, and especially among black youth. In

1981 the unemployment rate in Brent was over 10 per cent, but the number of unemployed young black people, as a proportion of total unemployed youth (under 25 years) was over 42 per cent, the highest in London.

This must also be seen in the light of the fact that Brent had the largest proportion of black and other ethnic minority people in Britain, with over 32 per cent of the population in 1977-1978 being non-white (Greater London Council 1983). Moreover the black population in Brent was concentrated in particular areas, with Stonebridge estate being one such area with over 70 per cent black residents. Thus de-industrialization was at the centre of the urban crisis of the late 1970s and the 1980s, with its effects predominantly being visited on particular groups and particular areas.

Moreover, the changes in nature of capitalist production affected labour not only in the form of job loss. Changes in employment patterns had occurred, with increasing shift work, casual work, part-time work, temporary work and short-term contracts, sub-contracting and 'sweating'. Employers were now seeking greater flexibility in their relationship with the labour market in order to meet fluctuations in demand or changes in the nature of demand. According to Atkinson (1985) a 'changing orthodoxy' in employment patterns could be identified, an 'emerging model' that

> is one of horizontal segmentation into a core workforce, which will conduct the organizations' key, firm specific activities, surrounded by a cluster of peripheral groups. Their twin purpose is to protect the core group from numerical employment fluctuations while conducting the host of non-specific and subsidiary activities which all organizations require and generate. (Atkinson 1985: 3)

Atkinson's model has been subject to criticism (Pollert 1988, 1991, Curry 1993), mainly in terms of whether or not employers are, in fact, engaging in flexible employment practices *strategically*, rather than reacting to their contemporary situation, and the extent to which the model proposed by Atkinson adequately described and explained employers' practices.

Nevertheless, the effects on local populations were to reduce even further the prospects of gaining regular employment on reasonably good pay, terms and conditions. Large sections of the unemployed faced not only difficulty in obtaining employment, but also the prospect that the only employment opportunities available will be in peripheral rather than core employment areas. Both the Bus Garage Project and the Charlton Training Centre were established to promote the employment prospects of people who were in such sections of the unemployed. Moreover both were established within the context of interventions by the state, both nationally and locally, to deal with the growing problem of unemployment. It is to this area we now turn.

State Interventions on Unemployment

As discussed in Chapter 6, unemployment had been an issue which the poverty programmes of the late 1960s and early 1970s had been concerned with, especially the Community Development Project (Loney 1983). However, as we have seen, the attempts by Community Development Project workers to relate urban poverty to structural aspects of capitalism and the state were disapproved of by government, and they were ended. The government took action in the early 1970s on rising youth unemployment through joint initiatives by the Department of Employment and the Industrial Training Boards in 1971, and by raising the school leaving age to 16 a year later. Such intervention however had only a temporary effect and by the mid-1970s unemployment, especially among school leavers, was rising rapidly, particularly after the economic crisis in 1976 and the Labour government's response. By this time the MSC had been established, and this agency was called upon by government to develop schemes to cope with the problem. We have already noted how this state agency has intervened in the interests of capital. We shall now examine further the processes by which it does this, particularly as they relate to the two organizations studied.

According to a former chairman, Sir Richard O'Brien, the MSC was set up, in 1973, for three main purposes:

- to provide continuity in national manpower programmes;
- to raise the priority given to manpower issues in companies, trade unions, industries and government;
- to associate interested parties in developing manpower policy. (Quoted in Society of Civil and Public Servants 1982)

The Greater London Training Board (Greater London Training Board 1983) argued that the MSC arose mainly as a compromise between opposing views of labour market policy and intervention. The Conservative Government elected in 1970 wished to abolish the Industrial Training Boards (ITBs). These were semi-autonomous, decentralized sector-based bodies, established under the provisions of the Industrial Training Act, 1964, with considerable powers especially through a system of levy on employers. The ITBs used the levies on employers to fund their operations and pay incentive grants for training in areas considered to be strategically important for their industrial sectors. Opposition by the trades union movement, educationalists and many employers and employer associations to such proposed abolition led to the MSC being set up, but at the same time the ITBs lost their independent funding, this now being provided through the MSC from the Exchequer.

The MSC, although 'tripartite', was set up within the Department of Employment and staffed by the civil service. If free market forces could not be allowed to operate, a Government controlled, centralized agency would be established. The MSC was a small department at first but soon grew in size and power as special measures were introduced to cope with rising unemployment, especially among school leavers. The special measures introduced by governments to deal with the mounting pressure because of increasing unemployment were developed and operated by the MSC, at first, through a Special Programmes Division. The schemes were intended to deal with two separate groups, unemployed school leavers and unemployed adults. In 1979-1980 special measures accounted for 29 per cent of MSC expenditure. By 1982-1983 they had increased in percentage terms to 55 per cent, with £720 million being spent. This was later to increase to over £1,280 million (62 per cent of total expenditure) by 1984-1985, as the full effects of expansion of the revised special measures worked their way through. But of even more significance than the sheer scale of the special measures was the form that they took.

Schemes for School Leavers

The main scheme for school leavers during the late 1970s and early 1980s was the Youth Opportunities Programme, providing work preparation and pre-vocational training. At the same time the 'Great Debate' (Chitty 1989, Simon 1991) was taking place in education, raising questions about how well, or badly, schools were preparing young people for working life. The underlying ideology was that unemployed school leavers lacked the skills needed to be attractive to employers, and so some 'transitional' programme was required to enable them to develop such skills. Programmes of 'life and social skills' and remedial teaching in literacy and numeracy were developed, and work experience on employers' premises was organized, so that participants would stand a better chance of obtaining employment. High rates of success (around 80 per cent) were claimed in the early stages but, as youth unemployment continued to rise, placement rates dropped to less than 25 per cent in 1981-1982.

Increasing criticism was being voiced about the schemes, as being 'cheap labour', masking the true rate of unemployment, policing the unemployed youth, and merely an extension of the social security system (Network Training Group 1983, Greater London Training Board 1983, Benn and Fairley 1986). As with the earlier Community Development Project, the underlying ideology was that unemployment was the result of individual characteristics, in this case a lack of appropriate education and training: such an ideology has been described as 'blaming the victim' (Benn and Fairley 1986: 9, see also Ryan 1976). In 1983 the government introduced the Youth Training Scheme, replacing the Youth Opportunities Programme (YOP).

This followed the publication by MSC of the 'New Training Initiative' in 1981, which had followed a period of consultation by MSC. The MSC had declared that

> Training is not given sufficient priority in Britain ... Not enough training is done and some that is done is misdirected and wasted. The time has come when we need to draw up and agree objectives to which all of us – especially Government, employers, trade unions and the education service – can work in the 1980s. (Manpower Services Commission 1981b)

The three objectives set by the New Training Initiative were:

- developing skill training for young people so that individuals could enter at different ages and be able to acquire agreed standards of skills;
- moving to a position where all young people under the age of 18 would have the opportunity to continue in full-time education, enter training, or have a period of planned work experience combining work-related training and education;
- opening widespread opportunities for adults to acquire, increase or update their skills and knowledge throughout their working lives.

Following development work by the MSC the government agreed to fund the Youth Training Scheme (YTS). This was hailed in various quarters as a breakthrough in training provision in Britain. It received endorsement from the TUC, as well as the CBI, the two main partners with government in the tripartite Manpower Services Commission. It was ostensibly to have overcome problems with Youth Opportunities Programme, especially the work experience on employers' premises element in that scheme. However considerable criticism was made in some quarters at the very beginning (e.g. Network Training Group 1983, Scofield, Preston and Jacques 1983, Greater London Training Board 1983), and criticism increased as time went by and problems emerged (St. John-Brooks 1985, Benn and Fairley 1986, Walker 1988). One of the challenges made was that the Conservative government, in establishing YTS, was not intending to promote high quality training but, on the contrary, to undermine existing standards of skills training. It was argued that the abolition of 16 Industrial Training Boards (ITBs) in 1981-1982 was a necessary precursor, as they might have resisted government and MSC attempts to lower standards which they had developed.

The ITBs were tripartite bodies with equal numbers of trade union and employer representatives on their Boards, as well as further education service representation. The hostility of the Conservative government to the ITBs was clear from the measures taken to abolish the majority of them. An MSC Review Body, with nominees from the Confederation of Business Industry and from the Trades Union Congress had advised their retention (Manpower Services Commission 1980). The Government then instructed the MSC to undertake a sector-by-sector review of industrial training (Manpower Services Commission 1981a), but in *not a single case* did the review team advise the closure of an ITB. Yet in November 1981, whilst the consultation on the New Training Initiative was still in progress, the Secretary of State for Employment, announced his intention to close 16 boards.

The Conservative Government had ensured that the Secretary of State would have such powers by specifically including such provision in the new Employment and Training Act of 1981 (previously, under the Employment and Training Act of 1973, an ITB could be abolished *only if* the MSC advised such action).

The 1981 Act also provided for a change in the system of funding for ITBs, so that operational expenditure was to be met from a retained portion of employer-levies, rather than central government funding. The overall level for both the levy and the amount which could be retained was to be set only where a majority of employer members of the Board agreed, and within limits set by the Act. The Greater London Training Board argued that the abolition of the majority of the ITBs and the transfer of the rest into effective employer control 'could weaken trades union influence and restore market conditions at the same time' (Greater London Training Board 1986: 442).

In 1980 the Central Policy Review Staff (1980) had argued that apprenticeship was 'a restrictive labour practice', and that the notion of 'skill' had no place in a modern economy. In this context YTS can be seen as a key part of the deskilling strategy required to support capital which was experiencing a crisis of profitability:

> In training policy terms the Government sees 'modernization' as a framework within which employers will be free to move away from expensive training in broad transferable skills (like apprenticeship) to state-subsidised training in narrow, task-specific 'competences', determined by the employer. (Greater London Training Board 1984a: 13)

Schemes for the Adult Unemployed

The Labour Government from 1974 to 1979 experienced major difficulties with high and rising adult unemployment as well as youth unemployment, difficulties exploited by the Conservative Party in the 1979 general election campaign with a slogan of 'Labour isn't Working' on an image of a long, snaking queue of, presumably, unemployed people. The Labour Government had instituted, through the MSC, a set of special employment measures to attempt to provide temporary employment (anticipating economic recovery to boost employment) and training in new skills for workers whose industries were in decline (Gregg 1990). At the time that the Conservative Government took office, the main schemes for the adult unemployed, the Job Creation Scheme and Special Temporary Employment Programme were fairly small, and these were replaced in 1980 by the Community Enterprise Programme. This was intended to provide temporary employment and training for long-term unemployed people on work which had some form of community benefit. In a full year up to 30,000 full-time places were to be made available, paying 'the rate for the job' up to a maximum £89 per week.

In 1982 the government were caught between the declared policy of tight restraint on public expenditure, and calls for action to be taken on the problem of rising unemployment, now reaching three million. In his budget speech the Chancellor of the Exchequer proposed to provide funds for a programme of 100,000 places which would pay unemployed people an extra £15 per week on top of their unemployment benefit in exchange for work on community projects. This was widely condemned '... as the first step towards requiring the unemployed to work for their benefit, and being a means to undermine wages and jobs in the public sector' (National Association of Teachers in Further and Higher Education and Association for Adult and Continuing Education 1983).

In the discussions which ensued the MSC drew up proposals for the Community Programme, which the government accepted. The Community Programme (CP) came into operation on 1 October 1982, although existing Community Enterprise Programme schemes were allowed to complete their contracted period. There were many significant differences between Community Enterprise Programme and Community Programme. The Community Enterprise Programme had provided a maximum wage of £89 per week; although the individual maximum was raised to £92.50 a maximum average wage in any scheme was set at £60 per week. The reductions took place despite evidence supplied by a number of groups active in the field of employment initiatives that most workers on £60 would be entitled to various statutory benefits which would bring them above that if they remained unemployed; the reduction in wages thus provided no net reduction in government spending (Network Training Group 1983). Such an act is understandable in the context of the declared view of the government that the unemployed had 'priced themselves out of a job'.

One effect was that schemes began to establish a mix of full-time and part-time jobs, to enable some workers to be paid over the maximum average. As we shall see, this had a major and damaging effect in the Bus Garage Project. Another effect was that some local authorities began to provide 'top up' funding to raise the average wage. However, such funding had to come out of the 'twopenny rate', i.e. the local authority could spend money under the powers of section 137 of the Local Government Act 1972, to the limit of the product of two pence in the pound of rates collected. We shall discuss this later in this chapter. Whilst overheads and other operating costs under Community Enterprise Programme schemes had been met through MSC funding to an agreed budget for each scheme, under Community Programme an upper limit of £440 per place per year was set, and proposed expenditure had to be agreed when applying for approval for a scheme. Training and education was to be paid for out of the £440 per place, and/or by deducting up to £10 per week from the 'wages element', thus reducing still further the average wage payable.

Many Community Enterprise Programme schemes had been involved with community action. Some were established as Unemployed Workers' Centres, and engaged in campaigning about unemployment. A number of community resource centres gained Community Enterprise Programme funding for workers, and printing facilities were provided for local groups. Under CP this was deemed to be 'political' and forbidden: 'Projects must not involve political objectives' (Manpower Services Commission 1982: section 3.k). *The Guardian* (26 May 1983) reported that a check had been ordered by MSC on TUC Centres for the Unemployed to see that no posters or pamphlets questioning government policy were being used. It quoted an internal memorandum: 'Sponsors are forbidden from taking part in activities that serve a political purpose or which are likely to bring the commission and the Community Programme into disrepute'. Earlier, the Sheffield Centre Against Unemployment had lost its funding from the MSC for its 'political activities': this was being 'represented on one or two marches – that sort of thing' (MSC spokesperson quoted in *Guardian* 23 March 1983). Thus, in developing the Community Programme, the state can be seen to have learned from the experiments of the 1970s. As *Gilding the Ghetto* argued:

> Keeping the initiative is essential to the state's success. ... The poverty initiatives were primarily experiments with and on the residents of the older industrial areas ... Above all they were experiments on behalf of the central and local state. (Community Development Project 1977: 57-9)

A sophisticated scheme was thus devised for reducing official levels of unemployment, at low cost, in a way that enabled work within the community (which might otherwise be perceived as the responsibility of the local authority) to be carried out whilst reducing rate support grant, and which lowered the wage expectations of the unemployed and those in low-paid public sector jobs. This was devised by those in managerial positions in the MSC, a branch of the Civil Service, and incorporated management and administrative procedures to ensure that few objections could effectively be raised. And by establishing the Programme through CP sponsors and managing agents, the actual schemes were run by 'subcontractors' from the state. Any problems within schemes would be the responsibility of the sponsors. Yet the areas in which sponsors could make decisions were severely constrained.

Adult Training: The SkillCentres

The Community Programme and its predecessors were temporary work schemes for unemployed adults. The MSC had also been involved in retraining for adults, through the Training Opportunities Scheme. This provided income support for participants, and payment of the costs of training, on programmes which could last up to one year. Participants were required to train and/or study full-time, but could leave employment voluntarily to do so. The scheme was originally intended to enable workers to change occupation, which was considered a feature which would become common in a changing economy. Although the scheme originally included the opportunity to undertake further and higher education provided it was vocationally-oriented, most training was for skilled manual work and was conducted by the MSC's own staff in SkillCentres. These were to come under threat when the Conservative government came to office in 1979, and a programme of rationalization and commercialization was eventually carried out.

The SkillCentres did not have an outstanding record for training. They typically ran courses of six months in length, compared with several years for apprenticeship schemes in the occupations covered. Many employers did not regard individuals who had been trained at SkillCentre as having sufficient skill to meet their requirements, especially as unemployment rose and more conventionally trained and experienced workers came onto the labour market. Moreover the operation of SkillCentres came under increasing criticism from women's groups, especially as sex discrimination legislation provided the stimulus for women to challenge the traditional demarcation of most skilled manual work as work which was only for men. Over 95 per cent of trainees at SkillCentres were male (Benn and Fairley 1986: 262), and both the practices and the structures of SkillCentres were regarded as sexist. The Greater London Training Board's analysis was that, in terms of practices

> the MSC has no policy to deal with the sexual harassment that women experience, whether it be the discomfiture of women as a minority in a masculinist environment or open sexual harassment. In terms of structure, for example, the inconvenient location of SkillCentres, their 8 am starts, their lack of child allowance and creches, their lack of positive action on outreach, their lack of taster courses to encourage women, all militate against the involvement in SkillCentre training. (Greater London Training Board 1984b)

In 1982 a number of SkillCentres were scheduled for closure, as a part of 'rationalization' programme; the Charlton SkillCentre was one of these. The Greater London Training Board argued that this was a short-sighted policy, reducing the capacity for training in recession which would result in a shortage of skills when the economy came out of recession. The closure went ahead despite the local campaign to keep it open, and the formation of the Charlton Training Consortium arose from that campaign. In 1983 the operation of SkillCentres was taken out of the Training Division within MSC, and given to a newly created Skills Training Agency which was required to be self financing. To do this, training would be provided at commercial rates to employers for their employees, and training for the unemployed would be 'bought' by the MSC under a programme to replace TOPS. Training for participants on Community Programme schemes would be treated the same as employees of commercial firms.

The replacement programme for TOPS was proposed in the MSC's discussion document 'Towards an Adult Training Strategy'. This argued that the immediate emphasis in adult training should be economic, suggesting that 'more effort should be put into training and re-training those *already in employment or about to start a new job*, rather than into purely speculative training or training purely for stock' (Manpower Services Commission 1983: para. 25, emphasis added). The proposals for training for unemployed adults were to be 'more closely tied to good local employment prospects' and be concentrated on work preparation. This was to include '... work preparation linked to training in limited occupational skills for a specific job' (Manpower Services Commission 1983: para. 40).

Thus the MSC's strategy for the unemployed can be seen to link with managerial strategies for creating flexibility in the use of labour, the unemployed to be accustomed to low pay, temporary, part-time work, and if trained at all to be trained only in limited skills tied to short-term needs of employers.

Phoney Decentralization?

Following the abolition of most of the ITBs and the planned introduction of YTS and CP an *appearance* of decentralization was created by the development of Area Manpower Boards (AMBs) to oversee these programmes. These boards had employer, trade union, educationalist representation, as had the tri-partite ITBs; the AMBs also had voluntary sector representation. However the *real* control lay with the MSC and, through it, the government and the interests of 'free enterprise' which it openly championed. This was effected by incorporating trade unions in a way which took advantage of a key weakness in trade union organization. Trades unions are organized nationally and by industrial/occupational sector. Local geographical organization is weak with regard to national issues; trades councils (local associations of trades unions) were ignored in the development of the Area Manpower Boards. The boundaries for the areas covered by AMBs were the MSC's administrative boundaries, not those of local government. This resulted in local education authority (LEA) representation being limited because many LEAs were covered by any single AMB, and in the case of the Inner London Education Authority (ILEA)[1] (which was Labour-controlled) the AMB structure ignored its existence.

London was divided into four AMBs with the ILEA area divided between those four Boards. The role of AMBs soon became a limited one, restricted to marginal modification of proposals brought by MSC officers. Thus when in 1984 Fife AMB refused to allow YTS schemes which planned to use a private training organization to provide the off-job education element, whilst local education authority provision was under-utilized (seeing this as privatization of education), it was overruled by the MSC. In the same year the MSC Head Office overruled those AMBs which resisted the instruction to advise on the reallocation of YTS provision to cut the proportion of places on schemes not run on employers' premises: local officers were instructed to effect the cuts. Moreover the conditions under which particular projects could be set up were very tightly laid down, and local officers would scrutinize a proposal thoroughly before seeking AMB approval. Thus Area

1 The Inner London Education Authority was itself abolished in 1990.

Manpower Boards members were largely reliant upon information supplied by MSC staff which serviced the AMBs, and local direction control was a façade. From these various aspects of MSC policy and strategy we can thus see that the state has been concerned to intervene in the relations of production in the interests of capital. Indeed, the MSC had soon gained the 'title' of 'Ministry of Social Control' by critics (e.g. Benn and Fairley 1986).

However the effects were more than to produce an appearance of dealing with problems of unemployment and to contain unrest. The policies and actions were more than responses to the results of the crisis in capitalist production; they were positive attempts to assist capital to regain profitability by restructuring the nature of the relationship of labour to capital. In doing so various organizations which ostensibly sought to promote the interests of labour became caught up in the contradictions inherent in cooperating with such a state agency, to the detriment of their own declared policies. These include trade unions and the TUC, Labour controlled local authorities, and local community organizations. We shall later analyse the effects on the organization and management of the Bus Garage Project and the Charlton Training Centre, which in different ways developed within this context.

Race

Both the Bus Garage Project and the Charlton Training Centre were concerned with issues of discrimination and disadvantage experienced by black people. As such they can only be properly understood in the context of the riots which took place in 1980 (St Paul's, Bristol), and 1981 (Brixton, St Paul's, Toxteth, Chapeltown, Handsworth), and the sporadic outbreaks which occurred after those. The riots were characterized by:

- the kinds of areas in which they occurred: areas of chronic deprivation in terms of housing standards, jobs, and community facilities, and areas of high crime rate especially involving young people;
- the high proportion of young people involved, both black and white but especially black;
- the direction of anger and violence at a police force perceived as racist and repressive;
- the immediate and harsh response by the police, using specially trained and equipped squads (Scarman 1982, Kettle and Hodges 1982).

Investigations after the riots, including the 'official' investigation by Lord Scarman of those in Brixton, pointed to the deprivation and the predominant modes of policing as being the main contributory factors.

> The disorders were communal disturbances arising from a complex political, social and economic situation, which is not special to Brixton ... There was a strong racial element in the disorders; but they were not a race riot ... The riots were essentially an outburst of anger and resentment by young black people against the police. (Scarman 1982: 77-78)

The riots however did not 'just happen'. The historical development of the factors which led to the riots has been examined (see e.g. Sivandan 1982, Kettle and Hodges 1982, Benyon 1984) in terms of the development of capitalism, in particular during the boom of the 1950s and early 1960s. It was during this period that immigration from the 'New Commonwealth' took place on a large scale, to provide desperately needed labour in particular in lower paid, low skill, low status jobs (see also Moore 1975). Sivanandan (1976) draws parallels with the influx to continental Europe of 'guest-workers', migrant workers who were single units of labour. Such 'ready-made' workers save the country of immigration the costs involved in reproduction i.e. maternity and health care, education and so on (Gorz 1970). Black workers immigrating to Britain came at first as single units of labour, without families, which also meant that country saved in terms of 'social capital', infra-structural provision such as schools, hospitals, housing, transport. However the economic advantages went to capital rather than to improvement of social conditions of the general population.

In particular the shortage of housing forced the immigrant workers into the deprived and decaying inner city areas, and the resulting overcrowding increased their image of undesirability:

> the forced concentration of immigrants in the deprived and decaying areas of the big cities high-lighted (and reinforced) existing social deprivation; racism defined them as its cause. To put it crudely, the economic profit from immigration had gone to capital, the social cost had gone to labour, but the resulting conflict between the two had been mediated by a common 'ideology' of racism. (Sivandan 1976)

Of course, in Britain the nature of immigration was different from that in other European countries because of the nationality status which arose from the recent colonial past and the creation of the Commonwealth. Dependents joined immigrant workers, and the second generation began to be raised in Britain. Simultaneously the social provisions, housing, education, and so on, were arranged in such a way that semi-ghettoes were formed. As urban areas began to be redeveloped, mass housing projects using new and untried methods were favoured by local authorities which had, during the 1960s, introduced 'corporate management' approaches and in which there had been a growth in professionalized bureaucracy. Schooling for black children developed into a system which had racism structured in by regarding West Indian modes of speaking as 'bad English', by a curriculum which denied any value to the historical and cultural heritage of such children, by assigning confused and/or rebellious children to 'remedial' classes (Gillborn 2005).

The interim report of a committee of inquiry set up by the Labour Government in 1979 recognized the effects of racism on under-achievement, because of the stereotyping of 'West Indian' children as problematic, difficult or slow (Rampton 1981). Police and judicial systems operated in such a way that a significant proportion of a growing generation were criminalized (Moore 1975, Sivandan 1982, Kettle and Hodges 1982, Benyon 1984). Sivanandan argues that changes in the immigration laws had the effect of limiting immigration so that would-be immigrants became migrants, subject to similar controls as for migrant workers in other European countries. The state then needed to count the cost of racial friction in social and political terms. So anti-discriminatory legislation was introduced 'not to chastise the wicked or to effect justice for the blacks but to educate 'the lesser capitalists in the way of enlightened capitalism'. When despite some 'success' with first generation immigrants, especially in creating a 'black bourgeoisie' to which the state could hand over control of black dissidents, the 'second generation' showed signs of discontent more laws were enacted. Their purpose was to enforce change.

> For, the anxiety of the state about rebellious black youth stems not from the rhetoric of professional black militants ... but from the fear of the mass politics that it may generate in the black under-class ... and perhaps in the working class as a whole – particularly in a time of massive unemployment and urban decay. (Sivandan 1982: 123)

The distribution of the black population in Britain was not an even one. According to the Labour Force Survey in 1981, of those living in England and Wales whose origins were the West Indies or Guyana, 57 per cent lived in London. The relevant figure for those whose origins were the Indian sub-continent is 34 per cent. Moreover the concentration of the black population in London varied from borough to borough. Brent had the highest concentration with 33.5 per cent according to the 1981 Census (although Brent Council later claimed the 1986 figure to be 56 per cent), the highest concentration in Britain. Even then, there are major differences between areas within boroughs. Thus in the Stonebridge estate in Brent, the Afro-Caribbean population was over 70 per cent of the population of the estate. In Greenwich, with the smallest proportion for any inner London borough with 7.7 per cent, there are significant concentrations in particular areas, especially among Asian communities for whom English is not the first language.

Stonebridge

The Stonebridge estate is situated near to the major industrial estate of Park Royal, which had developed between the First and Second World Wars, following on the establishment of a munitions factory in the First World War. Other nearby areas in Willesden and Cricklewood also had been major centres for employment. The structural changes referred to earlier particularly affected the area. Phizacklea and Miles (1980) undertook a study of the position of black migrant labour in the political process with field work in this area (referred to by the name of the old borough, Willesden, which was merged with the Borough of Wembley to form the London Borough of Brent in 1965). Examining the history of the area they argued that it was a declining centre of capitalist production, differing from the process of industrial development and decline occurring in other parts of Britain only in the fact that the cycle had a short time-span. Firms had moved out of the area because of restrictions on their capacity to expand, especially as national and international rationalization of capital took place requiring larger units of production. Much of the vacated floor space was taken over by firms in the distributive and service sectors, such as wholesale and retail cash-and-carry centres which also required much land for vehicle parking. Such firms required fewer workers and what jobs were created tended to be low skilled and low paid. Phizacklea and Miles point out that, at the time of their study, there had been a shortage of skilled labour in the borough, which may have been due in part to the policy of the GLC and the local borough councils to encourage labour to move to New Towns and Development Areas. Other studies of this policy had indicated that it was the younger, skilled workers who moved, and Phizacklea and Miles conclude that the majority of redundancies are likely to have been of semi-skilled and unskilled workers.

The effects of such changes can be seen in the circumstances of black people living in Stonebridge and neighbouring areas. By 1981 the unemployment rate in Stonebridge estate was 17.5 per cent (compared with 10.1 per cent for Brent as a whole), but for young people in the 16-19 age group the rate was estimated to be over 50 per cent. For Brent as a whole it was estimated that young black persons were four times as likely to be unemployed as young white persons (source: HPCC Bus Garage Project Steering Group 1984: 11-12). Many young black people became involved in crime, in particular street robbery and burglary. The Council's Stonebridge Victim Survey, 1983, concluded that 'the problems in Stonebridge are less of a primarily policing concern and more to do with Housing, Social Services and Leisure concerns. High unemployment is of crucial importance' (cited in HPCC Bus Garage Project Steering Group 1984: 11).

The use of marijuana was common and supplying it was a source of income for many, despite its illegality. As in other areas of high Afro-Caribbean population, the possession, use and supplying of marijuana was a source of friction between them and the police. Thus for many young black people in the area criminality was a normal part of life. Three of the four founders of the Community Council had served prison sentences. The ingredients of riots in other areas were present in this area (see above discussion of Scarman Report on the Brixton Riot), and in July 1981 a group of youths did in fact begin to gather and march towards the nearby shopping area of Harlesden intending to riot. They were dissuaded by two members of the group who later formed the Community Council (Williams 1992).

The development of that group, and the Bus Garage Project must be understood within this context, including, as we shall see, the fact that no riot took place.

Charlton Training Centre

The immediate context of the establishment of the Charlton Training Centre is somewhat different. Here the aim of improving the employment prospects of black people was more generalized, with the Centre having a wider catchment area rather than being intended to provide for the needs of an identifiable local black population. Moreover, the positive action programme was itself more general, being intended to tackle the disadvantage experienced by black people, by women, and by gay men and lesbians, as identifiable groups within the working class. In fact the origins of Centre lie more clearly in campaigns led by women's groups, concerned to change the low level of participation by women in SkillCentre training (Wickham 1985, Marsh 1986). The Greenwich Employment Resource Unit was actively involved in such campaign work, and the particular worker concerned was one of the main initiators of the short-lived campaign to reverse MSC plans to close the Charlton SkillCentre and thus of the emergent Consortium that would manage the Centre.

The separate provision of training for women was an early proposal in the development of the Consortium's plans. However the proposals on meeting the needs particularly of black people was also an early one, especially as representatives of local black community groups became involved. The involvement of the GLC, which was carrying out the London Labour Party's manifesto commitments on meeting the needs of black people and of women, provided a fundamental thrust to both aspects of positive action, but without any unifying philosophy. This we shall see when analysing the events which developed, especially in terms of the tensions between various elements of the positive action programme.

The Local State and the Community

Local Authority Funding

Both the Bus Garage Project and the Charlton Training Centre can not be properly understood without taking into account significant changes in the arena of the local state (Cockburn 1977). Both organizations were established through large-scale funding from local authorities and had continuing involvement of those local authorities in their development; however, they were also established as 'community initiatives' and not as directly under local authority direction and control. This must therefore be examined in terms of the significant trends in the relationships between local authorities and community and voluntary groups.

First of all it is important to remember that local authorities are restricted by law with regard to the activities for which they may provide funding. Their powers are established by Acts of Parliament, and there are certain duties that are imposed on them. Thus they are required to provide certain services, such as education and housing for homeless persons, and have a 'fiduciary duty' to ratepayers[2] to balance fairly the interests of those ratepayers with the interests of persons who are expected to benefit from any funded activity. The expected benefits of any expenditure must be commensurate with that expenditure, and financial prudence is always an overriding requirement.

The courts have intervened on many occasions where disaffected persons have sought to oppose the actions of local councils. In the 1920s the councillors of the London Borough of Poplar were prevented from paying their staff what they considered a 'living wage' because the court ruled that the council was able to employ such staff on lower wages. In 1981 the London Borough of Bromley brought a High Court action against the GLC on the reductions in London Transport fares. The case eventually went to the House of Lords where the GLC was overruled. The Audit Commission has powers to audit local authority expenditure and judge on whether certain aspects of activity exceed the powers of the local authority. In the mid-1980s, Councillors in Lambeth and Liverpool were surcharged and disqualified for deciding to delay setting a rate within a 'reasonable' time, when those councillors sought to persuade central government to increase their allocation of rate support grant.

So local authorities were (and still are) severely constrained in their ability to provide funding to local community organizations. Local authorities are empowered under local government legislation, originally section 137 of the Local Government Act 1972, modified by subsequent legislation, to engage in activities which are not covered by other aspects of local government statutes. This power was used, during the 1980s, especially by Labour controlled authorities that have sought to implement policies for which other local government statutes provide no powers. The total amount of expenditure was (and is), however, restricted (originally to income from rates to the value of the product of twopence in the pound). This was not an extra source of funding but provided for a portion of rate income to be used for such 'extra' activities. Thus when a council decide to use such powers in effect it has to do so by foregoing the use of such income for other activity which it was empowered to engage in through other statutes. Moreover the limited amount available under the 'twopenny rate' may potentially led to tensions when demand exceeds supply and where the particular use of the finance was viewed by many local people as a 'waste' and as inflating the level of rates imposed.

Urban Regeneration Funding

One source of extra funds for local authorities, during the period of the two projects studied, was the Urban Programme, a scheme intended to promote urban

2 Now Council Tax payers.

regeneration. This had been established by the Local Government Grants (Social Needs) Act 1969, and expanded by the Inner Urban Areas Act 1978. The Urban Programme provided central government funds to add to local authority funds for innovative projects aimed at combating economic, environmental and social problems in urban areas. A local authority seeking such funding provide one quarter of the cost, the (then) Department of the Environment providing the rest. The authority may fund its own projects or projects by voluntary and community organizations, but each project was subject to approval by the Department. In 1983-84 the grants provided totalled £288 million, of which about £50 million went to voluntary and community organizations.

The Urban Programme was administered through three systems of funding. *Traditional Urban Programme* was the remnant of the original programme, and was allocated on a project by project basis to any local authority which had pockets of 'special social need', and there was a maximum limit set for any project. *Inner City Programme* was established under the 1978 Act and provided for two forms of 'special' relationship between central and local government. *Partnership Programme Authorities* received specific allocations of funds for each financial year, and the Department of the Environment participated in the development of the councils' policy on tackling the needs of the areas covered. The second form of special relationship applied to *Inner City Programme Authorities* which also received specific allocations and, additionally were required to prepare a coherent programme of projects which was subject to Department of the Environment's approval.

There were no limits set for projects under Inner City Programme funding. The London Borough of Brent was given Inner City Programme status in 1983; Greenwich has access only to Traditional Urban Programme (Stewart and Whitting 1983, National Council for Voluntary Organisations 1985). Urban Programme funding was identified as the major source of public funding for the capital funding for converting the Stonebridge bus depot, and provided the funds to lease the site of the Charlton SkillCentre.

Stewart and Whitting (1983) examined the operation of the Urban Programme with respect to funding for ethnic minority projects, pointing out that this was marginal, and arguing for more attention being paid to the main programmes of government to tackle racism, discrimination and disadvantage which 'reflect ideologies and values in British society' (Stewart and Whitting 1983: 5). Nevertheless they argue that its most important attribute was the potential it provided to black groups for increased identity, self help and involvement with the political process. In examining the workings of the Programme in practice they pointed out that, as well as many well established, experienced, well resourced groups, there are also many smaller, younger, relatively ill-organized and under-resourced groups. Such groups often find difficulty in preparing applications, and the sources of assistance are often not apparent to them. The processes by which funding decisions are made tend to result in groups without a record of project management faring less well than those with such a record. This particularly occurs when criteria are not explicit and traditional criteria are relied upon.

> These rate public accountability highly and lay stress on the organizational and managerial competence of groups. Based heavily on personal knowledge of groups, on traditional approaches to voluntary action, on predominantly white cultural values, these (normally implicit) criteria represent a risk avoidance strategy which inhibits innovation in general and black innovation in particular. (Stewart and Whitting 1983: 33, emphasis added)

Problems also arose in implementation. Stewart and Whitting point to the problem of staffing which was a sensitive issue in the voluntary sector because of the widely held belief that having paid workers was not in the spirit of voluntarism. This resulted in tensions within projects when staff feel aggrieved that their pay and conditions of service are not commensurate with comparable posts in local authorities. Moreover difficulties arose as projects develop because the initial staffing arrangements, including job descriptions and pay scales, may not be appropriate to changing conditions. The emphasis on managerial capability in grant approval was carried through into implementation, with the requirements to account for expenditure of grant according to recognized professional standards. There are cultural differences between a traditional local authority service department and a community group which may be organized and run as a cooperative. Moreover many groups were, and still are, funded by more than one agency; in the 1980s, the Manpower Services Commission was a major agency providing such funding. The Conservative government had put pressure on local authorities to submit 'capital schemes' (buildings and equipment) rather than 'revenue schemes' (covering wages and running costs) (National Council for Voluntary Organizations 1985). So funding for wages and running costs has begun to come more and more from MSC through the Community Programme, as described earlier.

Stewart and Whitting (1983: 39) state that 'our evidence supports the Home Office research based suggestions that the plurality of grant-giving bodies and the interactive nature of funding produces a wasteful and inefficient system'. They point to different approaches to supervision and liaison adopted by the various funding agencies, and different procedures for this. The MSC was particularly regarded as providing for the 'most difficult relationship' with the Urban Programme. Referring to the MSC's particular view of its role in funding schemes which were related to labour market policies, they argue, based on their study of the experience of several schemes, 'the aims and criteria of the MSC may well conflict with those of a self-help oriented voluntary organization which has the individual or the community as its main focus' (Stewart and Whitting 1983: 41).

In analysing the events in the Bus Garage Project we shall see that such multiple funding, particularly the involvement of the MSC, had significant detrimental effects on the management of projects, effects which were in the interests of maintaining the capitalist nature of the state.

Greater London Council

The election of the Labour Party to control of the Greater London Council in 1981, the policies and strategies it developed and implemented, and its abolition in 1986 after a major campaign for its retention, is a highly significant chapter in the history of local government in Britain. The period of administration by the Labour Party was presented by the Chair of the Industry and Employment Committee as 'heralding a new kind of municipal socialism' (Ward 1983: 24). The Labour-controlled GLC received both acclaim and attacks for doggedly attempting to carry through the commitments of the manifesto on which it campaigned in the 1981 GLC election. That manifesto, over 100 pages long, spelt out in detail a variety of policy proposals and the means by which they would be accomplished. The commitment to a policy of improvements to public transport and cheap fares probably received most publicity.

However there were many other policy aims of the GLC's ruling party: anti-racism, equal opportunities for all irrespective of race, gender or sexual orientation, retention and expansion of employment, support for cooperative and community enterprise, support for trades unions, retention and expansion of quality training provision, improvements to the provision of housing. Many of these policy aims overlapped; for example section 11 of the manifesto (on employment and training) indicated how transport and housing policies were involved in carrying through a manpower plan for London. The policy aims also had implications for the style of local government adopted by the ruling Labour group. New standing committees of the Council were set up, for example the Ethnic Minorities Committee, the Women's Committee, and the Greater London Manpower Board, later renamed 'Greater London Training Board'. New departments were created and staffed, mainly through external recruitment, as 'Support Units' to these new committees. The Greater London Enterprise Board was set up as a separately incorporated body. Appointments to senior ranks of the Council's officer establishment were made much more on the basis of the appointees' understanding of, and support for, the policies. A much greater emphasis was put on consultation and involvement of the local people of London in carrying through the policies, especially through support for voluntary and community groups. In the foreword to a guide on GLC grant aid to such groups, the Council leader Ken Livingstone wrote:

> Our policies have been directed towards a more open and participatory form of government and we have established a close relationship with London's voluntary sector as part of this strategy. … GLC has taken the voluntary sector seriously and given it a full partnership role in the provision of London's services. (Greater London Council 1984: Foreword)

The Bus Garage Project certainly fitted with much of these policies, being a local black community initiative concerned with job creation. The GLC provided almost half of the capital funding for acquiring the site and indicated willingness to provide about half a million pounds towards conversion costs of the sports area. The post of Business and Cooperative Development Coordinator was funded by the GLC, together with a grant to cover some of the running costs of that section. When the Project was seeking 'top-up' funding for workers on the Community Programme scheme, Brent Council obtained 'unofficial' agreement that the GLC would provide it, and cover the money provided for this purpose by Brent Council in advance of the relevant GLC Committee meeting.

The origins of the Charlton Training Consortium also fitted in well with the policy aims. It began as an opposition to MSC policy and action, thus gaining Greater London Manpower Board support. It was intended to be a broadly based consortium of local community groups developing its programme in collaboration with the community, thus gaining the support of Popular Planning Unit. It aimed to link training with job creation especially through worker cooperatives, thus gaining support of the Greater London Enterprise Board and the Industry and Employment Committee. It aimed to provide training primarily for those most discriminated against in public sector provision and in the labour market, women, black and other ethnic minority people, disabled people, thus gaining the support of the Women's Committee and of the Ethnic Minorities Committee.

However, in addition to the legal constraints on the GLC as on any local authority, the Labour administration was attempting to promote its radical policies from within an established local government structure. Michael Ward, Chair of Industry and Employment Committee, describes some of the difficulties this presented and the actions taken by the Labour group to overcome resistance from GLC officers (Ward 1985). Moreover there are inevitably tensions involved in the attempt to carry through relatively clearly developed policies within a participatory approach. As the Capital and Class Editorial Collective points out, the Left did not win control of the GLC 'as a consequence of an upsurge in the strength and confidence of the Labour movement. Labour's (precarious) success was the result mainly of Tory unpopularity' (Capital and Class Editorial Collective 1982: 312).

The coincidence of perceptions and intentions of the members of a community organization with those of the ruling Labour group on the GLC (even assuming total coherence between different policy aims and the strategies for implementation) could not be assumed. Any tendency for a community organization seeking GLC support to propose or take actions contrary to GLC policy aims caused tensions in terms of the response made by the GLC. The stated commitment to a 'participatory form of government' and to giving the voluntary sector 'a full partnership role' precludes an authoritarian response. Inevitably, the GLC would either have to have refused or withdrawn public support for the community organization, or if it decided to fund it there would be a tendency for key members and officers of the GLC to attempt to influence the direction being taken by the organization. Either way there were significant implications for the community organization which may have sought GLC funding *particularly because* other sources of funding were refused by other agencies, or because other agencies would impose conditions which are unacceptable to the community group. We can see this in both the Bus Garage

Project and the Charlton Training Centre. The delay in granting 'top-up' funding for MSC-funded workers in the Bus Garage Project was the immediate cause of the dispute which took place, and the non-renewal of the funding after it was eventually granted affected the employment strategies adopted. The involvement of the GLC as a corporate member of the Charlton Training Consortium, and of an officer of the Greater London Training Board, was a continuing locus of influence over the direction taken.

Summary

Thus when we examine the context of the establishment and development of both organizations we can identify major trends in the nature of the relationship between capital and labour, as capital attempted to restructure both nationally and internationally. This had an impact on the nature of state activity as the rate of unemployment rose sharply, especially among particular groups. Interventions by the state may be characterized not only by repressive methods of containing discontent, but also by attempts to alter the nature of the labour market to enhance structural changes in employers' manpower policies. The state's interventions to deal with the problems which arose from the disadvantage and deprivation of black people must be situated in the history of immigration and of the place of black workers and their families in the changing labour market conditions over the past three decades. The relationship between the central and the local state provided the immediate context for both organizations examined, especially in respect of the emergent patterns of funding for local groups. Central government, under Conservative control, has been concerned to control the nature of such funding, especially to conform to its declared commitment to aiding capitalist enterprise. Labour controlled GLC funding, whilst based on policies which directly oppose Conservative government policies, nevertheless have presented constraints on the directions taken by community organizations.

Chapter 8
Processes of Domination in the Management of a Job Creation Project

From Job Creation to Employment Dispute

The feasibility report on the Stonebridge Bus Garage Project, presented in December 1981, stated that philosophy of the proposals was that the local community wished to 'improve local conditions through their own efforts and in their own way' (Stonebridge Bus Depot Steering Group 1981: 4). In line with this a job creation scheme was established in spring 1983, under a Government funding programme, to provide jobs for local people who had been unemployed for long periods. Jobs were created and workers recruited in office work, site security, and building work, the latter enabling disused parts of the buildings to be rehabilitated for short-term occupation pending start on full-scale redevelopment. The job creation scheme was expanded a few weeks later, with extra building workers being taken on. Most workers were under 25 years old, Afro-Caribbean black, lived within the local area, and had been out of work for at least six months, in most cases never having held jobs on a continuous basis.

However within a few days of the start of the job creation scheme, relationships between building workers and the supervisor and scheme manager became tense. In early July, four weeks after the scheme was expanded, a major dispute occurred with the Project Coordinator issuing suspension notices to a large number of building workers, notices which they refused to accept. A scheme which had been established to provide opportunities for local disadvantaged black young people had become, for them, an experience which reflected the conditions in society at large: oppressive, dominating, uncaring, authoritarian, unwilling to hear their demands for fairness.

The Community Council which had emerged from within the local black community, led by the peers of those who were now in dispute, held seven of the 12 posts in the Steering Group, and so were formally the top management of the Project. How had they ended up in such a situation, where they appeared to be oppressing the very people they claimed to represent?

Events Leading up to the Dispute

Becoming an Employer

Among the immediate tasks facing the Steering Group for the Stonebridge Bus Garage Project was the negotiation for funding for employees. During the summer and autumn of 1982 applications were successfully made to central government departments, the Greater London Council, and the European Social Fund, to be able to recruit a team of 'coordinators'. These were people with relevant qualifications and experience to lead the development work in particular areas: architectural planning, financial management, business development, training, as well as overall project coordination and administration. By December 1982 these posts had been filled and detailed work was started on obtaining more funding for the support staff for these areas and for short-term building work to be able to use the semi-derelict site. The Steering Group decided to 'take advantage of' the Manpower Services Commission's (MSC) Community Programme. This was a job creation scheme intended 'to help those who have been out of work for some time and to help local communities' (Manpower Services Commission 1982: 107).

The Steering Group as a 'sponsor' was approved by the MSC to start a project, receiving MSC funds under the Community Programme, whereby a number of unemployed, mainly local people were employed to work in aspects of the Bus Depot Project's work: site security, administration, clearing out and demolishing certain areas of the depot, rehabilitating some semi-derelict rooms to bring them into temporary use as offices and so on, and decorating areas redeveloped by outside contractors. The decision taken by the Steering Group to become a Community Programme sponsor was based on the idea of meeting a number of aims:

- creating jobs for local people who had been unemployed for a long time, in many cases since leaving school;
- enabling these people to obtain training in order to improve future job prospects;
- involving local people in the development of the Project, rather than 'importing' outsiders through outside contractors;
- bringing certain areas into use on a temporary basis (up to three years) whilst waiting for complete redevelopment;
- reducing the overall capital funding required by capitalizing revenue funding i.e. wages for building workers paid by MSC is converted into capital in the form of buildings and so on.

The MSC approved the start of the scheme in April 1983, with an initial building work team of a scheme manager, a senior supervisor, a supervisor and six workers. A further team of a supervisor and six workers was approved at the same time, to start a few weeks later. Subsequent approval was obtained to take on a further two supervisors and 15 workers to start at the same time as the second team. A management committee was set up by the Steering Group to deal with the operation of the scheme. The committee consisted of seven persons:

- two members of the Steering Group who were nominated by the community council;
- the two local borough councillors, nominated by the two main political parties, who served on the Steering Group;
- two of the 'professional' workers employed by the Steering Group viz. the Project Coordinator and the Training Coordinator;
- a representative of the local trades council.

Meetings of the committee were held frequently at first to deal with a number of policy issues including terms and conditions of employment, discipline and grievance procedures, interview and selection panels. The hours of work were originally set at 37½, that is, 9am to 5pm with half-hour lunch break. This was the pattern for existing employees. The disciplinary and grievance procedure adopted was similar to that which applied to existing employees. Matters of discipline or grievance that were not resolved between an individual and her/his immediate supervisor would be referred to the Scheme Manager. If the matter were not resolved by him it was referred to a disciplinary and grievance panel consisting of three members of the management committee convened by the Project Coordinator. If the matter were still not resolved it would be taken to the full management committee for the final decision. Moreover any recommendation for dismissal of an employee had to be taken to the management committee who had sole authority to dismiss.

Disciplinary Action

At the first meeting of the management committee after the Scheme Manager had been appointed he recommended that the hours of work be changed to 39½ hours, that is, 8am to 4.30pm (4pm on Friday), which was agreed. The subsequent meeting held three weeks later was fairly short and dealt only with minor items. A meeting was held five weeks after that at which the Scheme Manager reported that three workers were giving unsatisfactory performance despite disciplinary action being taken by the supervisor and himself. The Project Coordinator was instructed to convene a disciplinary and grievance panel meeting. The meeting of the panel, the first such meeting, was held a few days later. The members were the two local councillors and the Project Coordinator. Before dealing with the disciplinary matters the Training Coordinator joined them to outline some guidelines on dealing with disciplinary and grievance issues.

During the discussion the councillors disagreed between themselves on certain points and one left accusing the other of having made up his mind and being determined to sack the individuals concerned. The meeting was abandoned without dealing with the disciplinary cases brought. A further meeting was convened with two other members of the committee in place of the councillors. Of the three workers concerned only one turned up. The panel, after hearing the matter, gave a formal warning to the individual and put him under 'review' for a month. The panel agreed that letters be sent to the two absent workers instructing them to appear before the full management committee at its next meeting, with the warning that failure to do so would be taken to be their resignation.

At that next meeting of the management committee one of the workers turned up. The Scheme Manager made complaints about the individual's poor timekeeping and his 'poor attitude' to work. When challenged by the Training Coordinator to explain what this meant, the Scheme Manager stated that he did not intend to 'bandy words' and that it was not normal practice to challenge a manager's judgement. The meeting chairman intervened to insist that the Scheme Manager provide the explanation which the committee required. After hearing the individual's comments the committee decided to act as the disciplinary and grievance panel had done for the first worker, that is, formal warning and put under 'review'.

At the same meeting it was reported that the Scheme Manager had taken disciplinary action with all three workers concerned by reducing their wages. The rate of pay originally agreed was £92.50 per week. The MSC paid £60 of this, the maximum average wage payable from MSC funds under Community Programme rules. The other £32.50 was obtained in the form of a grant from the local authority. The Scheme Manager had reduced the three workers wages to £60, a decision he had taken in collaboration with the Project Coordinator. After discussion it was decided to uphold this action, but that in future such action should only be taken following the decision of the disciplinary and grievance panel.

The Scheme Manager then reported that an additional 21 workers and two supervisors had been recruited with MSC approval (as described earlier). The Training Coordinator expressed concern at this decision, which had been taken without reference to the management committee, or the Steering Group, or the Employment Sub-Group of the Steering Group. The concerns expressed were:

- no funds were available to top up the wages of these extra workers so they could only be paid £60 per week;
- the 21 workers consisted of six workers who could be employed for 43 weeks, and 15 workers who would be employed for only 13 weeks;
- the terms and conditions of employment needed adjusting for these additional workers.

The Project Coordinator then reported that an application had been put into the Greater London Council for a grant to provide 'top-up' for the extra workers. The committee decided that the Project Coordinator should meet with the employment sub-group to examine all these issues and report back to the next committee meeting. At the next meeting the Scheme Manager asked that the full disciplinary and grievance procedures come into effect *only after* an individual had completed a four week probationary period, that is, that *he* should have the right to dismiss someone *he* deemed unsatisfactory within the four week period. After discussion in which the Training Coordinator disagreed with the request, the committee granted it, subject to the individual's right of appeal to the committee. The committee also agreed that during the probationary employees would only be paid £60 per week. After that period, if kept on, they would be paid £92.50, the difference being backdated, subject to 'top up' funding being obtained.

At the following meeting the Scheme Manager queried the decision taken regarding wages. He stated that he felt he ought to be authorized to grade workers according to 'their ability to attend on time and their willingness to take instructions'. He explained a grading system with a four point scale: 6, 7, 8, 9, which related to different rates of pay: £60, £70, £80, £92.50 respectively. The committee rejected this and endorsed its previous decision that workers who were retained after the probationary period would be paid £92.50 backdated to their starting date. That same meeting was asked to hear the appeal of a worker who had been dismissed after three weeks, for leaving work without good reason or permission. The appeal took over an hour to hear during which time the Scheme Manager and supervisor made certain statements which many members of the committee understood as indicating that they would resign if the worker were to be re-instated. The committee decided by a 4:2 majority to endorse the Scheme Manager's decision to dismiss.

The Dispute

The following day the Project received news that the Greater London Council committee responsible for deciding on the application for the grant to top up wages had not, in fact, considered the matter at its meeting earlier that week. The next meeting would not be held for a further three months and so there was no top up funding available from that source. After consulting the Project Coordinator, the Scheme Manager called all the building workers together at the end of the morning shift to explain the situation and to tell them that, unless anyone had other suggestions he intended to keep all workers on their existing rates of pay. After lunch the workers held a mass meeting to discuss the issue. The Scheme Manager returned from his lunch break to find the mass meeting in progress and instructed them to be as quick as possible and return to work. The Project Coordinator left for his lunch break and when he returned found the meeting still in progress; by now it had taken one hour. He told them they had no right to hold a meeting in work time and instructed them to return to work within five minutes. He returned in five minutes, took the names of all present and informed them that they were suspended pending hearings by the disciplinary and grievance panel 'on which he intended to sit'. He then had letters to this effect typed up which he personally posted on his way to a meeting at the town hall later that afternoon.

Some workers left the Project site, but others refused to go, insisting that they wanted to see someone else about the matter. The Training Coordinator returned at about 4.30pm from an all-day visit away from the Project. On hearing of the dispute he asked members of the Steering Group who were on site to join him in meeting with the workers. At the meeting the workers agreed to their chosen representatives meeting with a group selected by the Project chairman to represent the Steering Group, in order to examine ways of settling the dispute. Meanwhile they agreed to return to work.

The 'negotiating teams' met the following day. The workers' representatives listed their grievances which included:

- that they had been promised that their wages would rise to £92.50 after the four weeks probationary period;
- that workers arriving more than a few minutes late were being sent home with no pay;
- that they were being given no information about the Project except what the Scheme Manager gave them, and they did not trust this;
- that their views were being ignored;
- that most had not received written statement of the terms and conditions of employment, particularly in respect of the length of their contracts.

At a subsequent meeting the two parties reached agreement on wages, fixed penalties for lateness, and representation of workers at Steering Group meetings.

Following this, relationships between workers and the Scheme Manager remained strained. Disciplinary matters continued to be brought to the grievance and disciplinary panel, in most cases resulting in warnings. Some workers sought assistance from the local, trades council-sponsored Unemployed Workers' Centre to organized workers at the Project in a union. Initially there appeared to be considerable interest, but this soon waned, and recruitment efforts foundered as workers expressed the view that the unions could do very little for them as wages were set by the MSC.

Analysis

In seeking to understand and explain how and why these events unfolded as they did, two points are worth observing at the outset. First, the organization is not a commercial enterprise which has to generate profit in order to survive in the marketplace. At first sight there is no imperative to generate surplus value through exploitation of labour, and so we might not expect the strategies adopted by management to have similarities with those adopted by management concerned to extract surplus value. Secondly, although the Project was ostensibly set up as a community initiative and with social aims, the events described are clearly not examples of those social aims being fulfilled. The explanation therefore needs to address the fact that a situation of industrial conflict should arise in a project of this nature.

We noted earlier, in examining the context in which the Bus Garage Project was established, a number of key features:

- the local area had suffered from the industrial restructuring, with massive job losses;
- the Stonebridge estate had a very high proportion of black residents, with the estate developing the characteristics of a semi-ghetto similar to other inner city areas in which rioting took place;
- the state had been developing its interventions into the workings of the labour market with temporary work and training schemes which were concerned to enhance the structural changes taking place to provide flexibility for capitalist enterprise;
- although the Labour controlled Greater London Council was committed to various policy initiatives intended to promote the interests of labour, and various groups within the working class including black people, there were various tensions and conflicts which arose in implementation of these policies.

Within this context we may see that the dispute which arose through setting up a Community Programme scheme was the result of the managerial strategies to control labour and of the rather ill-organized resistance to these by workers.

Three main themes emerge:

- the dominance of the aims of production over the social aims of the Project;
- managerial strategies to maximize discretion over the productive process;
- managerial strategies to maximize power to reward or punish workers.

Aims of Production

The dominance of the aims of production (building work) over the social aims (employment and training) is evident in a number of events:

- working hours were increased;
- disciplinary action taken because workers were late, absent or 'not working as hard' as the Scheme Manager or supervisor deemed appropriate;
- extra workers were taken on at lower pay rates, in order to do more building work.

The question of whether the building programme established by the Scheme Manager was realistic was never seriously raised. But it certainly seems questionable whether low paid, unskilled people with limited experience of regular employment and the normal habits and disciplines of work would be able to become sufficiently skilled, motivated, or self-controlled (socialized in terms of timekeeping and responding to supervision), or be able to sustain physical effort for long periods. No systematic attempt was made to help workers adjust to the rigours of work. No systematic attempt was made to involve workers in decision making to ensure that their concerns were taken into account.

The processes by which this came about can be traced to the pressures on both the Community Council and the local authority. The Community Council wanted to get work started as quickly as possible and to involve local people in the Project by creating jobs. The local authority, faced with cuts in rate-support grant, pressure from ratepayers objecting to rates rises, and other community groups demanding grants, wanted to minimize funds given to the Project. The Community Programme seemed to enable this twin set of aims to be met but at the cost of working within parameters set by MSC. Caught up in the desire to get building work carried out, the Management Committee responded to problems and needs for decisions by putting the building work before social aims.

Discretion Over the Productive Process

The Scheme Manager and original supervisor both came from traditional organizations, in both private and public sector. Their position in all this can be understood in terms of the processes by which capitalism ensures the reproduction of agents who work in the interests of capital, although not being themselves members of the capitalist class. These processes include economic (differentials in wage rates) and ideological (training and socialization) (see Therborn 1982). Understandably they were concerned to maximize production and in order to do this they needed to maximize discretion over the productive process. This was effected, with some resistance from the workers, by a variety of strategies. Labour mobility was one main approach: workers were moved from job to job with no regard to their skill development; jobs were fragmented and individuals given small parts to carry out; workers changed jobs often, except where the job required limited skill, such as basic demolition; when the team was expanded workers were moved between supervisors. Although this strategy appears to be opposite to the process of specialization (division of labour), often cited as a key aspect of the capitalist mode of production, it is in reality part of the general strategy of obtaining flexibility by limiting the skills of such workers.

Discretion over the productive process was also effected by taking on extra workers without Committee approval, and then withholding written terms and conditions. This was a clear strategy for selecting longer-term employees from a larger short-term supply of labour. The concentration of valid information in the hands of the Scheme Manager denied the workers any chance of identifying where they could gain a degree of control over the productive process. Workers were never given information on building plans, and since they were moved from job to job, were unable to establish any degree of discretion over the work.

Power to Reward and Punish

Linked to these strategies for maximizing discretion over the work process were strategies for maximizing managerial power to reward or punish. One such strategy, which was later stopped by the Management Committee, was to link wages to a managerially-defined performance appraisal. It was the approaches to discipline (punishment) which illustrates clearly the strategies. At first the Scheme Manager attempted to deal with disciplinary matters (as defined by himself) without reference to the disciplinary and grievance panel. This included using non-approved sanctions, such as reducing wages by a third. After a period of failure to 'remedy' problems the panel was invoked quite suddenly. At the first meeting of the panel the Scheme Manager attempted to use vague judgements of the 'offender' and expressed resentment at challenges to his judgement *and at his right, as the manager, to make such judgement.*

Although this strategy was unsuccessful (in that the worker was retained) a subsequent strategy was successfully introduced. This was the suspension of the full disciplinary and grievance procedures during the probationary period, giving the Scheme Manager complete discretion to dismiss during that period. At the first (and only) appeal against dismissal pressure was exerted by the Scheme Manager and supervisor to ensure that their decision was upheld. Finally the fact that during the period that the scheme has operated only one grievance matter was dealt with by the panel gives an indication of where the power over reward and punishment developed most.

The analysis of events within one organization can thus be seen to illustrate how the interests of capital are carried through from the wider economic and social totality. While not fitting in directly with the Braverman-model of deskilling, the events in this organization illustrate the general managerial strategy of control over labour. Moreover, because this has taken place within what was intended to be an attempt to create an alternative to capitalist enterprise, the result has been not only to contain discontent but also to reproduce the conditions for the continued reproduction of capitalism.

Chapter 9
Community Initiative or State Enterprise? Setting up Stonebridge

In this chapter, we step back from the level of a specific episode discussed in previous chapter to examine the history of the Bus Garage Project from 1981, when the Community Council was formed, until mid-1985 when agreement was reached with central and local government to implement plans for the rebuilding of the bus depot. We can identify three main phases which the Project went through:

- initial development and formulation of plans;
- intensification of problems leading to a crisis, in which the future of the Project was under threat;
- development of a set of changes to the building plans and to the organization and management of the Project, which once agreed enabled the Project to continue towards the redevelopment of the site.

We shall examine this history through the three phases identifying significant events and trends, then analyse the organizational and management issues which arise. In doing so, we shall be able to identify key elements in the transition of what was originally intended to be a community project into what may be better described as a 'state enterprise'.

Phase One: Management of the Project

Purchasing the Site

After Brent Council approved the Project Report and purchased the bus depot a new Steering Group was set up whose composition was originally planned as

- seven members nominated by the Community Council;
- three members nominated by the local Neighbourhood Forum;
- two local councillors, one from each of the major political parties.

The local Neighbourhood Forum was an umbrella organization bringing together the various local self-help groups, and serviced by Brent Council staff – run Joint Neighbourhood Project. This was an initiative by Brent Council to provide a team of workers from the different departments of Brent Council (housing, social services, youth and community service, education and so on) to be based on the estate to provide outreach to the community. Until early summer 1984, only one representative from the Neighbourhood Forum attended meetings. The Community Council therefore not only had a formal majority on the Steering Group but also an effective majority of seven to three.

Brent Council purchased the bus depot site in March 1982 and granted a licence[1] to the Steering Group to develop the site. During the summer and autumn of that year approaches were made to a variety of public authorities for funding for operating costs, including wages. It was planned that a number of professionally qualified and experienced staff be recruited to enable the Project to be set up and progressed and also to train particular individuals as their assistants. These assistants would, it was hoped, take on the full running of the Project at an appropriate stage.

The 'professional' posts identified as needed were:

- Project Coordinator;
- Finance Coordinator;
- Cooperative/Business Development Coordinator;
- Training Coordinator;
- Legal Advisor (half-time);
- Building Project Coordinator;
- Administrator.

At this stage it was planned that the first three mentioned coordinators would have two assistants each, and the Training Coordinator would have one assistant. These positions were to be filled by six of the seven Community Council representatives on the Steering Group, and one (the assistant to the Training Coordinator) by another member of the Community Council. The seventh representative on the Steering Group was by now being paid by Brent Council's Youth and Community Service, and was the chairman of the Steering Group. Funding was obtained for each of the 'professional' posts and for the assistants to the Finance Coordinator and the Cooperative/Business Development Coordinator. Advertisements were placed and each post filled during November and December 1982.

Getting Started

During the first three months of 1983 further funding was sought under the Manpower Services Commission's Community Programme for the operative staff deemed necessary and for the supervisors of these staff. In addition a manager of

---

1 Granting a licence provided the Project Steering Group with the legal right to use and engage in activities on the premises, on conditions set by the Council, but without the legal right of occupation that a lease would provide. This clearly placed the Project in a state of uncertainty and dependence.

the Community Programme scheme was to be employed. The application to the Manpower Services Commission (MSC) was approved and staff employed. In planning the staffing structure it was decided that the two assistants to the Project Coordinator should be funded as supervisors under the Community Programme scheme, one for administration the other for security.

From the description of the structure of the Project so far it will be noticed that the Community Council was heavily involved in the management of the Project. Not only did many of the members act as members of the Steering Group, they also were assistants to the 'professional' coordinators. The stated rationale for this was that the Project was to be run by the local community, particularly by the black community group that was the main initiator of the Project. Insofar as the individuals involved lacked the experience and expertise to manage the detailed work needing to be carried out, they were to be trained on the job by the 'professionals'; and insofar as there was some concern that the involvement of 'professionals' might lead to the Project diverting from the ideas and aspirations of the community, the community would control them as their direct employer. The Steering Group held meetings every week on Wednesday mornings, attended by the six assistants who were members of the Steering Group, the other Steering Group members, and frequently by the 'professional' staff. Steering Group meetings were also attended each week by the Policy Coordinator (i.e. the assistant to the Chief Executive) of the local council. In April 1983 the Steering Group was formally incorporated as a company limited by guarantee, with the members of the Steering Group becoming the members and directors. From this point on the meetings began to be referred to as 'Board meetings' or more commonly 'BoD [board of directors] meetings'.

During the first few months of 1983 meetings of the staff were held every Monday morning to discuss the internal issues of the management of the Project. These were convened by the Project Coordinator and usually held in one of the portacabin offices, with people crowding into what limited space there was, positioning themselves where they could around the office furniture. The Community Council held its own meetings usually on a Monday evening. In May these arrangements changed so that the Community Council meetings were held on a Monday morning and the staff meetings at lunchtime on the same day. However the Community Council meetings usually over-ran the time allotted and members arrived late for the staff meetings.

The large increase in staffing in April had already resulted in attendance at staff meetings being restricted to 'professional' staff and their assistants, the MSC Scheme Manager, the Chairman and another member of the Community Council who was employed as administrative assistant to the MSC Scheme Manager. The position of the assistants to the Project Coordinator posed a problem for the formal description of the management structure of the Project. The arrangement for funding their wages, as supervisors of specified work groups within the MSC Community Programme scheme, was a deliberate one. However the rules of the Community Programme state '[d]irectors of companies which sponsor Community

Programme projects may not be employees of any project sponsored by their company' (Manpower Services Commission 1982: para. 17).

To circumvent the problems this would create, because the Community Council wished these two members to remain as Directors, a 'MSC Scheme Management Committee' was set up to deal with all matters relating to the scheme. This committee consisted of:

- the two local Council nominated directors;
- two Community Council nominated directors;
- the Project Coordinator;
- the Training Coordinator;
- a representative of the local trades council.

After seeking and obtaining confirmation in writing that the two directors/supervisors would not be involved in the management of the Scheme, the local MSC officer responsible for overseeing the Scheme agreed that they could be appointed. During the period before 'professional' staff had been recruited the Steering Group had been assisted by two members of the local Cooperative Development Agency (CDA) one of whom took on the half-time Legal Advisor position, working on other CDA activities the rest of the time. Various sub-groups were set up to deal with the details of decisions and action needed. These included

- the Building Sub-Group;
- the Finance Sub-Group;
- the Employment Sub-Group.

After the 'professional' staff were appointed the Finance Sub-Group was effectively replaced by the Finance Coordinator's department; the Employment Sub-Group virtually ceased to function. The Building Sub-Group however continued to operate with the involvement of the second member of the local Cooperative Development Agency, who had experience in building projects. His involvement with the Project became markedly less after the firm of architects for the second phase of redevelopment was chosen by the Sub-Group. The firm preferred by the Building Sub-Group was not the choice preferred by the CDA advisor, who tried to persuade the Steering Group to overturn the decision of the Sub-Group, but without success.

The Building Sub-Group met almost every week on Tuesday mornings and reported to the Steering Group the next day. Most reports were given orally and were information-giving rather than decision-requesting. The Sub-Group originally consisted of:

- the Chairman;
- the Finance Coordinator;
- the CDA advisor;

- an assistant to the Cooperative/Business Development Coordinator (who was also one of the Community Council directors).

To this Sub-Group were later added the Architect and the replacement Building Project Coordinator. Reports to the Steering Group were normally given by the assistant to the Cooperative/Business Development Coordinator, even after the new Building Project Coordinator was appointed. The recruitment of 'professional' coordinators was not a smooth affair. The first choice for the Cooperative/Business Development Coordinator decided not to take up the post and the post was re-advertised. The Building Project Coordinator post was taken by an architect from the local council's Development Department on a secondment. However a number of issues regarding reporting and accountability were not settled, and the individual obtained and took up a post with a firm of consultant architects. The post remained vacant for several months with the CDA advisor acting as temporary Building Project Coordinator until the split over the appointment of the architects. Eventually, after the job description was altered to remove the requirement that the occupant be a qualified architect, a local person known to the members of the Community Council was offered and took up the post without it being re-advertised.

Building Plans

The Architect began to develop a scheme design in consultation with various individuals, as directed, regarding the requirements for the community centre. The Architect presented sketch designs at a Steering Group meeting on 22nd June 1983. At that meeting the local council's Policy Coordinator who attended every Steering Group meeting expressed concern that the estimated overall cost was now greatly in excess of the original estimate presented in the proposals agreed by the Council. The original estimate had been £1.8 million; the new estimate was £3.5 million. In answer to the suggestion that the alternative of demolishing the existing building and building anew might be less costly the Architect stated that in his view this would not be so. It was agreed that the Steering Group would review certain areas of costs and look for alternative ideas, and that the Building Sub-Group should look at the phasing of the building to reduce costs or delay work to give time to raise more funds.

The Architect was instructed to proceed with drawing up the Scheme Design, which was subsequently approved along with the plan to carry the redevelopment work in three stages. The staging would, it was considered, enable the building work to be carried out over three financial years to match the provision of funding from public authorities. The arguments put forward to justify the increase in the estimated overall cost were summarized in a report sent to the Department of Environment dated September 1983. The arguments included:

- a longer useful life for the building than originally assumed;

- a higher standard of building to boost community morale;
- reduction in running costs because of higher standard of building;
- increase of floor space (almost doubled).

Funding Problems

At this time attention became focused on work being undertaken by the Finance Coordination department on income and expenditure projections for the community centre. A report was presented to Brent Council's Policy and Resources Committee in early November 1983 but was not well received. In particular the estimates of income were considered to be very over-optimistic. It was agreed at the next meeting of the Steering Group that the 'Space–Income' report be revised after consultation with various relevant officers of the local Council. The revised report was examined by those officers and subjected a number of criticisms including:

- running costs were underestimated;
- proposed charges would be too high for local people;
- the income estimates were over-optimistic during the initial period of operation.

By now the estimated total cost for the centre, including fees and equipment and so on, was put at £6 million.

The Policy Coordinator, in reporting the views of the Council's officers drew attention to the fact that only £1.5 million could be assumed from Urban Programme funding, and so a funding gap of £4.5 million remained. At the Steering Group meeting of 12 October it was noted that the Quantity Surveyor's estimate of the cost of the first stage was now £1.4 million, excluding equipment. Work continued on the preparations for the invitation for tenders for the first stage of the redevelopment work. Invitations to tender were issued in January 1984 although the funding presumed to be available, all from public authorities, was only about £1.2 million. The tenders made all exceeded the Quantity Surveyor's estimate, the lowest being over £1.87 million. The Architects were then instructed to develop a set of proposals for reducing the costs.

The first set of proposals presented on 22 February still required more funding than that presumed to be available. Moreover the Policy Coordinator pointed out that the Department of the Environment and the Greater London Council had not yet approved the funding they were expected to provide (£1 million). The Steering Group therefore agreed not to let any part of the stage A contract until both sets of funding were approved. A meeting was arranged by the Policy Coordinator between himself, certain Steering Group representatives and senior officials of the Department of the Environment. At the Steering Group meeting on 28 March, the day before the arranged meeting the Policy Coordinator announced that the meeting would also be attended by representatives of the Greater London Council, the MSC and the Sports Council (who were expected to provide about £100,000).

When the members of the Steering Group queried this he replied that matters would be 'simplified if all funding bodies were present, to iron out all points/ problems together'. Also at this Steering Group meeting the Policy Coordinator reported that the Department of the Environment was considering commissioning an independent consultant's study of the Project.

Staffing Problems

The Project experienced difficulties over staffing from the beginning of the period in which people were employed. As stated earlier the recruitment of two of the 'professional' workers, the Cooperative/Business Development Coordinator and the Building Project Coordinator, caused problems. The Building Project Coordinator position took four months to be filled. This was partly due to a redefinition of the job, and the post was eventually filled by an individual who was known to the Community Council members. This person played a limited role in key decision making, most reports on the Building Sub-Group's work being given by another member of that Sub-Group with the Building Project Coordinator being absent from the meeting. The problems experienced in using the MSC's Community Programme have already been discussed. The most serious problem, however, was the position of Project Coordinator. It became clear fairly early in 1983 that this individual was not carrying out his role to the satisfaction of the Steering Group or that of many Project staff. His involvement at Steering Group meetings was a limited one. The action delegated to him was the result of proposals from members of the Steering Group, the Policy Coordinator, or other 'professional' staff. Similar situations arose in MSC Scheme Management Committee meetings. The incident in which he took it upon himself to suspend the whole of the building team (discussed in previous chapter) caused considerable loss of 'face' among all staff on the Project.

By 26 October 1983 the Steering Group was openly critical of the Project Coordinator's performance. A number of items of work which were deemed to be part of his role were not being carried out, or were unsatisfactory. Steering Group meetings, coordination of which previous meetings had explicitly stated to be his responsibility, were confused. Other 'professional' staff were complaining that their own workload was increasing and problems being created for them by what they saw as the poor performance of the Project Coordinator. On 8 November the Community Council asked him to resign but he refused. He did not turn up for a conference of 'professional' staff and Community Council members held the following weekend to review relationships between these groups. In February 1984 a formal review panel was set up to deal with the problem. The Project Coordinator was given a three month period in which to improve his performance. Although the situation did not seem to improve the Steering Group waited for the end of the three month period before convening another review panel meeting. The meeting was postponed because the Project Coordinator was absent through illness. Before the meeting was rescheduled to take place the Project Coordinator and the Steering Group negotiated an arrangement whereby he would leave at the end of July 1984 but work only part-time during June and July.

By now the Consultant's study was underway, and the Project had no person employed who could be regarded as the main point of formal contact between the Project and the public authorities. To fill this gap the Steering Group, acting on the recommendation of the Policy Coordinator, decided to appoint the assistant Project Coordinator as Acting Project Coordinator pending the outcome of the study and any recommendations regarding management structure and staffing.

Phase Two: The Crisis

Consultant's Study

At the Steering Group meeting on 4 March 1984 it was reported that the Department of the Environment had decided to commission a study by consultants (plural) and a draft set of terms of reference was presented to the Steering Group by the Policy Coordinator. Some minor amendments were made and the draft sent to the Department of the Environment and the Greater London Council. The Department replied with draft terms of reference. The points of particular interest in these include

- the heading was 'Terms of Reference for *independent* consultants' study' (emphasis added);
- reference was made to the idea that the Community Centre was intended not only to serve the local community but also the Afro-Caribbean community in the London area;
- the consultants were invited to conduct a financial appraisal of the Project;
- reference was made to the prospects of achieving a 'partnership development' with the private sector;
- the consultants were to consider whether the Project has an appropriate management structure and access to the professional advice it needs;
- the urgency of the study was emphasized.

The Steering Group considered the terms of reference and submitted amendments including

- that the study should be an economic and a social one as well as a financial appraisal;
- that the prospects of a partnership arrangement with the private sector should take into account 'the objective of ensuring direction and control by the local community';
- that as well as considering whether the Project has access to the professional advice it needed the consultants should examine whether it has access to the financial and other resources it needed.

The Department sent a copy of the final terms of reference which did not include any of the amendments proposed by the Steering Group but did include a new additional item, that the consultants be invited to consider 'the suitability of the proposed means of construction as to best meet the needs of the Project'. The letter accompanying the copy of the terms of reference stated that, unless there were objections, the Department proposed to appoint, as consultant (singular), an individual who was on secondment from his firm of Accountants to the department 'as a member of the professional team appraising applications for urban development grant ...'. The named individual [from here on, referred to as 'the Consultant'] started in late May and was expected to report in late June. He was given a desk in the Finance Department and began a series of meetings with various individuals in the Project and in the local Council. However after about a week he revealed to one of the 'professional' staff that he was attempting to develop a set of proposals that took the existing level of assumed funding from public sources as the basis for the cost of the Centre. At the Steering Group meeting on 30 May the Consultant was invited to describe his plans for conducting the study. When he began to do this several members of the meeting expressed concern over some of the points he made.

In particular, after revealing in a comment which appeared to be an 'aside' that he was not an independent consultant but a secondee to the Department, those present began to press both him and the Policy Coordinator on whether there was a set limit to the capital cost of the Project. The reply from the Policy Coordinator was that there was indeed a limit to the public funding presently being considered. Concern was also expressed that the Consultant seemed to be taking a very narrow view of the nature of the Project especially with regard to its potential as a centre providing facilities specially for the Afro-Caribbean community in the metropolitan area, and its place within the local economy. After this meeting the Consultant was seldom on site and postponed a number of meetings. Two weeks later the Steering Group formally noted that he had been absent for some time and that he had a number of meetings outstanding. Meanwhile the Minister dealing with Urban Programme projects wrote to the Chairman saying that the consultant group had 'found the project very exciting and are keen to make a contribution to seeing that we have a thoroughly well-based plan for the future *within the level of resources that we know can realistically be achieved*' (emphasis added).

The Consultant returned to work at the project about one week after the Steering Group had written to the Department. The reason given for his absence was that there had been urgent work at the Department. The Consultant asked for a meeting with the Steering Group and this was set for 9.30am on 4 July. It was decided that this meeting was to be for members of the Steering Group only. However the Consultant arrived about one hour later than the appointed time after a meeting at the town hall with the Policy Coordinator, the two local councillors on the Steering Group, and the Chief Executive. The Policy Coordinator came with the Consultant and stayed in the meeting. At the meeting the Consultant stated that if he had to make a final report at that time he would not be able to justify the expenditure of Urban Programme funds. He therefore wished to meet with the Minister to agree a longer period to do a more in depth study; meanwhile, he urged, the Steering Group should do more work on the income projections. Among the points he made were:

- that in his view the Steering Group had been badly served by its professional advisors (when pressed on this he replied that he meant the architects);
- that he could not justify the cost of the proposed scheme, which was now estimated to be liable to cost £5.8 million;
- that the income projections were unrealistic;
- that the Department could not justify spending Urban Programme funds on a restaurant which would be used by people who were not from the local neighbourhood.

The last-mentioned point was a reference to the plan that a restaurant be built in the Project which would specialize in Caribbean cuisine and draw its clientele from a wide geographical area.

During this period the Development Department of the local Council had been discussing with a private sector construction company the possibility of it taking over responsibility for the re-development of the site on a management contract basis. This company, part of a large multi-national group, had devised a modular system of building for sports and leisure centres using prefabricated materials. The Steering Group examined outline plans for such an approach which would have involved demolition of the existing building. After visiting an example of such a centre, and after considering the advantages and disadvantages, the Steering Group unanimously decided not to proceed on this alternative.

By now the Architect had developed, on the instructions of the Steering Group, a modified scheme design the cost of which would be between £2.3 million and £3.6 million. In mid-July at the suggestion of the Policy Coordinator the Steering Group agreed to prepare a report describing the background and development of the Project, current plans, and recommendations to various bodies who were being asked to make key decisions regarding the future of the Project. The report (HPCC Bus Garage Project Steering Group 1984) was prepared with the involvement not only of the 'professional' staff and their assistants, but also with help from the local Cooperative Development Agency and the Trades Council-sponsored Local Economy Research Unit. The preparation of the report was coordinated by the Policy Coordinator and the final report, approved by the Steering Group, sent to the various bodies including the Department of the Environment. The Consultant met with a group of Steering Group members and 'professional' staff on 31 August, after receiving the report. At this meeting he made a point of stressing that he was appointed to do the study as 'private sector secondee' to the Department, and that he was under the same constraints 'as any other civil servant'. He stated that he saw the Project going through three stages: the original scheme design, the questioning of this by officers of the local Council and other funding bodies, and now the latest set of proposals. He was faced with two different sets of proposals, the latter ones being totally different from those on which he had originally been instructed to report. He reminded those present that he had suggested that the Steering Group come up with proposals which were 'realistic ... less ambitious ... costing less ... within the finance that was remotely possible ...'. He criticized the latest proposals as being unrealistic because they were based on an assumed continuing subsidy from public sources of between £0.4 million and £0.7 million per year. He stated that this was equivalent to a capital investment of about £5 million.

The group from the Project pointed out that the latest report was the outcome of rethinking by the Steering Group after his discussions with them on 4 July. The report was now the final view of the Steering Group and the one on which the Department of the Environment and other funding bodies were being asked to decide. The Consultant was informed that although the architectural, financial and other details of the Steering Group's proposals had changed 'in terms of the needs and demands of the local community the proposal was still substantially the same as it ever was'. When persistently pressed about the meaning of the term 'realistic' in respect of funding, i.e. 'how much is the Department considering putting into the Project?', the Consultant stated that 'Government doesn't work like that'. The meeting concluded with the Consultant being clearly told that 'the Project has now issued a very clear proposal for the Community Complex and we now expect the Department of the Environment to respond to it'.

Consultant's Report and Reactions

The Consultant's report arrived at the Project three weeks later. It was highly critical of the Project, and the conclusions included

- that the Project had 'developed to such a size that it appears to have lost touch with the "grass roots" (local) community ...';
- that it was 'now time to take the project out of the hands of the professionals and return it to the community';
- that the Steering Group should 'consider the special needs of the people (of the area) ... consider their priorities ... in the light of existing facilities in (the borough)';
- that the Steering Group might need to consider its own composition;
- that the Steering Group should adopt 'a more rigorous approach' to excluding elements that are either not special to (the area), provided elsewhere (in the borough) or aimed at a larger area than that immediately surrounding (the neighbourhood);
- that after carrying out these actions the design of the centre 'must be realistically projected within budget'.

The report included a review of the Project's plans prepared to date. However the length of the report was only 19 pages many of which contained tables of figures, and others only one paragraph (because of the sectioning of the report). A few days later a Policy and Resources Urgency Sub-Committee was held by the local Council. The agenda included the Consultant's report and a proposed response written by the Policy Coordinator. The 'Summary' paragraph of the covering report (nominally from the Chief Executive) stated that, in his view, the Consultant's report was 'totally unsatisfactory and suffers from so many deficiencies that it should now be put to one side. What is required instead is a response to the report published by the … Steering Group in August 1984'.

The response to the report, agreed by the Sub-Committee, made many hard-hitting comments on the Consultant's approach and his report:

- 'It is inaccurate and misleading and fails to understand the key features of the … Project';
- 'The report is extremely misleading and in places appears to deliberately misuse statistics to give a false impression of the Project';
- 'Many of the Conclusions in the report are based on judgement rather than objective assessment';
- 'The report contains factual errors';
- the Consultant 'does not appear to understand the Project';
- the Consultant 'does not appear to have read the Steering Group's August 1984 report in sufficient detail';
- the Consultant 'does not seem to have devoted sufficient time to the study';
- the report 'in no way constitutes an adequate response to the Steering Group's August 1984 report'.

A few days previously the Minister had visited the local area and attended a meeting at the Project with members of the Steering Group, the leaders of the main political parties, and people from the Project itself. The leaders of the parties each impressed on the Minister their dissatisfaction with the report and the need to get a decision on funding as early as possible. The member of the Steering Group nominated as responsible for publicity wrote a letter the following day to the Minister asking that he set up a meeting with all the bodies involved to discuss the Steering Group's report. He replied that he saw the next step as a meeting between officials from the Department, the Project, and Brent Council to discuss the Consultant's report.

Phase Three: Resolution

Aftermath to the Consultant's Report

During September and early October 1984 considerable 'behind the scenes' discussions took place between the Community Council members and senior officers of Brent Council. They discussed the possibility of replacing the Project's architects with the Council's own Development Department. In this way, argued the Council's officers, confidence in the Project by the various funding agencies would be restored. At the Steering Group meeting on 18 October the Board agreed in principle to dismiss the architects. On 29 October a meeting was held at the offices of the Department of the Environment between members of the Steering Group and officers of the funding bodies. These consisted of:

- five senior officials of the Department;
- the Consultant;
- the Chief Executive and Policy Coordinator of Brent Council;
- a senior officer from the Greater London Council;
- a senior officer from the Sports Council;
- four Community Council members of the Steering Group.

The most senior official from the Department opened by stating that the Minister and other Ministers

> were very impressed by the Project and were keen to see it make progress. However [the Minister] was alarmed at the way the Project had grown and concerned about its future revenue implications; he felt the [Community Council] had not been given the best advice in the past. A more modest approach is required within realistic financial parameters.

In the ensuing discussion on what was meant by 'realistic financial parameters' the GLC officer stated that there was some uncertainty about the half million pounds expected from the GLC, especially because of its impending abolition. The Sports Council official also raised problems about the funding expected. The Department officials indicated that if Brent Council were to submit an application for Urban Programme funding this would receive 'sympathetic consideration', and that they would support funding proposals from the GLC. The Project's decision to dismiss the architects was 'welcomed' by the Department, the GLC and the Sports Council. This decision had been formally made by the Steering Group at its meeting on 24 October, along with the decision to appoint Brent Council's Development Department as architects to the Project. Brent Council had in fact anticipated this decision, on the basis of the 'in principle' decision at the previous meeting, and had issued an advertisement which appeared in the *Guardian* on 24 October for a 'Community Project Architect'. The grading of the post was that of Principal Officer grade 2A, a full grade above that of the original Building Project Coordinator and that of the other 'professionals' in the Project.

Although there was an inevitable lead time before an appointment was made to this post, the Development Department started work on the Project almost immediately. In fact, as the Department had been involved in assessing the earlier plans and had suggested some form of 'new build', some preliminary work had taken place. The 'new build' suggestion was raised again and alternative schemes to that had been examined earlier were re-investigated. In addition alternative approaches to management of the rebuilding were examined, including:

- traditional procurement through competitive tendering on the basis of a complete design with Bills of Quantities;
- 'design and build', in which one main contractor takes on the full design and construction contract, according to criteria established with the client;
- 'management fee', in which the main contractor works with the client in the design stage, then manages the rebuilding including dealing with competitive tenders and coordinating subcontractors on site, for which an overall percentage fee is paid.

These were examined during the period from November 1984 to March 1985, during which time negotiations with funding bodies continued. In addition, discussions took place with a private sector development firm on the possibility of it taking on the funding of the workshops area of the community centre. The proposal was that in return for funding this portion of the redevelopment the firm would be granted a lease at peppercorn rental, and would collect rent from the management body of the community centre. The estimated level of funding required was half a million pounds, which would enable public sector funding to be concentrated on other areas. In December 1984 Brent Council's Policy and Resources Committee approved in principle such an arrangement, enabling those involved in negotiations to move towards detailed proposals. By April 1985 detailed proposals were ready for approval by Brent Council and the other funding bodies. A scheme design estimated as costing £3 million pounds was proposed, this estimate covering the costs of works, fees and equipment.

The scheme design was for rehabilitation of the existing building, rather than demolition and new building. A 'management fee' contract was proposed on the grounds that this would combine benefits of speed with the retention of significant control by the Steering Group, and so had advantages over 'design and build' where the client tends to have less control. Moreover the costs were likely to be lower than traditional competitive tendering, and more predictable. The main sources of funding were to be

- Brent Council, £1.5 million through Urban Programme allocations over the three years from 1985-1986 to 1987-1988;
- GLC, £500,000;
- a private sector development firm, funding the workshops area, £650,000.

The proposals were put to Brent Council's Policy and Resources Committee in April 1985, along with other information and proposals concerning the Project, and were approved. In the following weeks the details were discussed with the Department of the Environment and after some delay approval was given to the plans. However the Department insisted that Brent Council be responsible for the redevelopment work, and Urban Programme funding was to be retained and managed by the Council. All dealings by the Department were to be solely with the Council. Later in 1985 work eventually started on redevelopment, and by mid-1986 such work was on schedule for completion in early 1987.

Financial Management Problems

In addition to the problem that the original scheme design had exceeded the level of funding that was immediately in prospect, further problems of financial management arose. The Project's auditors took four months to audit the accounts and qualified their report, dated 28 August 1984, with 'Notes Regarding Weaknesses in the System'. They pointed to

- problems arising because finance department staff do not always carry out properly the operations of the accounting system;
- problems in reconciling the MSC account, which MSC insisted must be a manual recording system, with the MSC section of the computerized accounts;
- errors in journal entries;
- lack of management controls in other areas of the Project.

These problems had been seen earlier, in the form of delays and inaccuracies in financial reports to the Steering Group, to Brent Council and the MSC. However the most serious problem which arose concerned the unauthorized use of Project funds for supporting a construction company established by the Community Council. The company collapsed without repaying the debt to the Project, and with other debtors including the Inland Revenue. When, in September 1984, this came to the attention of the Chief Executive of Brent Council a meeting was held between him, the Policy Coordinator, the Project's Financial Coordinator and the Building Project Coordinator (the last-mentioned was deputising for the Acting Project Coordinator, who, with other key members of the Community Council, was visiting the black community organization in Watts, Los Angeles, which was seen as providing a model for what the Community Council might be). At the meeting the Financial Coordinator agreed to be suspended on full pay while the matter was investigated by Brent Council's internal audit department.

The internal audit department's report was received in January 1985, and discussed at the Steering Group meeting on 16 January. The report indicated that the problem was not an easy one to unravel, especially as the construction company's books were not available. Interviews were held with directors of the Project, the Finance Coordinator and Brent Council's Policy Coordinator. Essentially the problem had arisen when the Finance Coordinator had agreed to the Steering Group Chairman's request to exceed the £10,000 limit set on advance monies to be provided to the construction company. This advance was originally agreed to provide working capital for the company to buy materials, for work to be undertaken at the Project. The company's debts were rising and the Community Council were concerned that the collapse of the company would reflect badly on the Project. Further advances were therefore made, but the information was not supplied to the Steering Group. By September 1984 the amount had risen to nearly £20,000, and at this point the Building Project Coordinator reported the matter to the Policy Coordinator who then reported to the Chief Executive.

The internal audit report made several recommendations which were accepted by the Steering Group. In addition the Steering Group issued a letter of reprimand to the Finance Coordinator and also to the Chairman for their failure to notify the Steering Group of their actions. Eventually, after being suspended on full pay for over three months, the Finance Coordinator returned to the Project.

Management of the Project

The report produced by the Steering Group in August 1984 (the 'August Report') included a section on the organization, management and staffing of the proposed community organization which would be responsible for the centre. The proposals included details of how the transition would be made from the existing Project, concerned with developing the site and the future activities, to the 'community organization' which would operate and manage the redeveloped community centre. The proposals for this 'community organization' had been drawn up several months earlier by a sub-group on which had sat two 'professionals' and three members of the Community Council and Steering Group. These had been issued in the form of a consultative document to various local organizations including Brent Council. The Council responded by stating that it wished to await the outcome of the Consultant's study before commenting. Although the proposals were included in the August Report, Brent Council again did not respond directly.

Instead, a set of less detailed proposals were included in the report adopted by the Policy and Resources Committee in April 1985. The Steering Group proposals for the structure of the community organization had been drawn up on the basis of four principles, outlined in the consultative document and the August Report:

- that the structure should promote and develop community spirit;
- that the structure should promote active participation by all members of the community;
- that the structure should provide for full accountability by those given authority to the community which gave that authority;
- that the structure should be stable and promote continuity and consistency.

It was proposed that the community organization be established as a company limited by guarantee, with membership being persons nominated by various relevant bodies. Some of these bodies would be organizations which existed independently of the community centre, including Brent Council, the Community

Council and various local community groups. However, it was also proposed that those people and groups who used and worked in the community centre should be represented. This would be accomplished by establishing five 'internal councils' which each brought together those involved in the particular areas covered. These were to be:

- 'Business and Commercial Council', linking those who were engaged in self-managed commercial enterprises;
- 'Social and Welfare Council', linking the projects and groups which were concerned with social and welfare issues, such as the Under-Fives Centre,
- the Information Technology Centre and the Youth Centre;
- 'Arts and Culture Council', linking groups involved in such activities as drama, music, dance, writing, art and crafts;
- 'Sports and Leisure Council', linking various groups involved in such activities;
- 'Workers' Council', linking all those who were employed at the community centre.

The nominees of all these external bodies and 'internal councils' would collectively form the General Council of the community organization, the supreme governing body. The General Council would elect a management committee to direct, manage and control the centre. It was intended that the various internal councils would exercise considerable autonomy in their respective areas of involvement, within policies agreed by the General Council as interpreted in concrete situations by the management committee. The staff of the centre would be in one of three departments: sports, arts and culture, and central management and support services. Sports department would provide the various services required by that aspect of the centre's operation; arts and culture department would organize the use of recording and rehearsal facilities and would run the night club and catering facilities.

Central management and support services would provide all the general services required including business and legal advice, and engage in publicity and fundraising activities. Each department would be headed by a departmental manager, and the centre would be managed on a day to day basis by the manager of the community centre. However it was also anticipated that the arts and culture manager and the sports manager would also be very much involved with the respective internal councils, so that the users of these facilities would be actively involved in determining policy and practice in their particular areas.

The staffing proposals in the August Report indicated that the wages bill would be almost a million pounds per year in the early stages. This included provision for locally recruited staff to be trained, and it was expected that some staff would move on after training and that after two or three years fewer staff would be required. Some of the staff costs would be met from the revenue generated by the centre's operations. Nevertheless, it was anticipated that by 1989 the centre would be requiring an annual subsidy from various public agencies of just under £400,000. At the meeting held at the Department of the Environment on 29 October it had been reported that the Minister was concerned at the implications for continuing public subsidy. Moreover some of the revenue estimates were considered to be unrealistic. During the following months, reassessment was made about the level of staffing and the report to the Policy and Resources Committee proposed a reduced number of staff. It also proposed lower grading and salaries for the senior staff of the centre, with the proposed salaries for the Centre Manager and the Assistant Managers being reduced by almost 10 per cent.

However, it was more immediate circumstances that resulted in major changes in the management of the Project. The Project Coordinator had left in July 1984, although as stated earlier there had been problems with his performance of that role. The Financial Coordinator had been suspended in September and the Business and Cooperative Coordinator resigned in November. The initial contract for the Training Coordinator was due to expire in December 1984, so this was renewed to enable the future management and staffing structure to be assessed and decided before appointing a successor. This was agreed to by the Training Coordinator on a half-time basis.

Attention was therefore being placed on the question of what form of managerial structure should be developed, given these changes and the original intention that the Project/community centre should eventually be managed by local people, preferably members of the Community Council. In January the Policy Coordinator prepared a report to the Steering Group on 'Review of Current Management and Staffing Issues'. He drew attention to what he perceived to be problems of management and staffing, some which had persisted for some time despite previous discussions and agreements on remedies. These included:

- lack of adequate management information; lack of distinction between the role of the Steering Group and the role of 'professionals';
- need to strengthen the policy making and review functions of the Steering Group;
- lack of confidence by the Steering Group that its decisions will be carried out, resulting in spending a lot of time progress chasing;
- an increasingly smaller group of members of the Steering Group being actively involved in decision making, neglecting the need to strengthen links with the outside community;
- increasing demotivation and demoralization by the staff of the Project.

On the question of the excessive time spent in progress chasing and dealing with administrative detail he stated :

> If the BoD [Board of Directors = Steering Group] were to appoint someone as overall manager of the project who was responsible to the BoD for implementing their decisions and managing all project staff, and if the BoD had confidence in

> this person, and if this person had adequate support from other staff, then perhaps the BoD would be more willing to leave that person to deal with many of the matters which it tries to deal with itself. Then the BoD could concentrate more on policy making and review and perhaps could meet fortnightly or monthly instead of weekly.

He proposed that the management structure proposed in the August Report be implemented as soon as possible, and suggested that the current Building Project Coordinator be made Acting Manager while the process of recruitment of an overall manager and for the other unfilled posts took place. The Steering Group discussed the report at its meeting on 6 February 1985, and was generally accepted. After further discussions in the following weeks it was agreed that advertisements be placed for the posts of Project Manager, Assistant Manager (Support Services), Assistant Manager (Development), and Training Coordinator. The Assistant Manager (Development) post in effect was a replacement for the Business and Cooperative Coordinator but the job description had been changed so that there was now an emphasis on developing the business side of the Project directly under the control of the Steering Group.

In addition the Finance Coordinator's contract was terminated at the end of March, and this post was also advertised. A 'Review Panel' was set up to handle the recruitment process, including assessing the competence of internal candidates from the Community Council. The five members of this panel included the Chairman of the Steering Group and Community Council and another Steering Group member nominated by the Neighbourhood Forum. The Conservative councillor on the Steering Group, the Policy Coordinator and an officer from the Personnel Department of Brent Council made up the rest of the panel. In addition two other senior officers from Brent Council and one from the GLC acted as advisors to the panel.

These changes and the procedure for recruitment was detailed in the report to the Policy and Resources Committee in April 1985 and approved. The report stated that '[m]any of the problems experienced by the Board of Directors have resulted from inadequate servicing by their professional staff and the new staffing structure ... directly addresses this problem'. After describing the changes the report continued:

> The main characteristics of the revised staffing structure are as follows:- (1) A hierarchical rather than a co-operative structure. (2) Clearer delineation of reporting links and responsibilities. (3) A distinction between the provision of support services on the one hand and the development of new activities on the other. (4) The creation of a Project Manager post to act as the main link between the Board and its staff. (5) The creation of a management team of key senior staff chaired by the Project Manager and serviced by an Administrator.

During the discussions on the revised management structure the Steering Group had agreed new salary scales. The report to the Policy and Resources Committee stated that

> These salary levels involve lowering the salaries for some posts below their job evaluated levels. The Board have decided to do this because many of their grants – for example MSC – are not based on the same job evaluation system as Brent. ... If the Board were to pay all staff job evaluated rates they would require a large increase ... in their grant from ...Brent. In order to avoid this the Board have lowered some senior staff salaries and introduced a pay freeze for others. Salaries levels will be reviewed each year in the light of performance and productivity as well as taking account of the Project's overall financial position.

A few weeks later the posts were all filled, with the Training Coordinator and Assistant Manager (Development) posts being taken by external candidates. The Acting Manager was appointed as Project Manager, the previous Assistant Project Coordinator became Assistant Manager (Support Services), and the Assistant Finance Coordinator became Finance Coordinator. This was the situation at the time that agreements were reached with the funding authorities on the plans for redevelopment.

Analysis

In attempting to provide an analysis of the events in the Project it is again important to recognize the context in which it was started. We noted earlier that the immediate context was characterized by the effects of restructuring by capital, which particularly affected the black population of the local area. Moreover the various interventions by the state, both nationally and locally, had developed since the period in ways which served to maintain and reproduce the conditions for continuation of capitalism. In examining the Bus Garage Project we shall identify the key ways in which this embryonic community initiative was progressively transformed into an enterprise which posed no threat to the existing order, and served to sustain it.

Economic Control and Patronage

First, we can see that the Project was always subject to the economic patronage of the state. The large scale funding for the Project was not available from any other sources, and even Brent Council needed central government funding support through the Urban Programme. This situation arose very early in the life of the Community Council. At the time of the first calls for the bus depot to be used for community activities these two parties had already been partners in activities within the local area. The local Council had set up the Joint Neighbourhood

Project providing a local base for Council officers involved in the various services run by the council. The Community Council began life in a small local community centre, and worked with the council's Youth and Community Service during the summer of 1981 helping with the summer programme of youth activities.

Thus from the beginning the Community Council had operated from a basis of resources being provided by local state agencies. The local Council provided facilities to run a small youth centre, with key members of the Community Council being paid on a sessional basis through the Council's Youth and Community Service. These key members were able to use the opportunity of having premises available for various activities to engage in small-scale commercial activities e.g., running discos, operating pool tables, selling drinks. The vision of the bus depot project was one of expanding the commercial activity onto a grander scale. But both the early economic activity and the planned future activity were to be based on the 'gratuitous' provision of economic facilities by the state.

This contrasts with the statement in the original project report to Brent Council and the section on funding:

> It should be made clear at the outset that the bus depot project is not proposing that the Council, Central Government and other agencies convert the Depot into a multi-million pound community centre and sports complex. (Stonebridge Bus Depot Steering Group 1981: 4)

The sources of funding quoted in section 5 of that report consisted of a long list of 'other agencies' (in addition to Brent Council) including:

- The United Nations;
- The European Economic Community (EEC)[2] and overseas governments;
- British Government departments (Department of the Environment, Department of Industry, Home Office, Department of Employment and Manpower Services Commission);
- the Sports Council, the Football Trust, and the Arts Council;
- charitable trusts;
- Prince of Wales Jubilee Trust;
- the private sector;
- Greater London Council.

The report also asked that an immediate grant of £5,000 be made to help the Community Council to carry out activities related to the Project during the next three months. So from this early stage there was an emphasis on obtaining economic resources from the state and from others sectors of society which are dominated by capitalist perspectives. The Project continued on this path. Except for a very small element of 'self-generated' income the running costs were from state sources: the

2 Forerunner to the European Community.

local Council, Urban Aid and Urban Programme, EEC, Home Office, but most especially from the Manpower Services Commission (MSC) under the Community Programme. This has resulted in the tendency for the Steering Group to frame its planning and decision making within the strictures of the various programmes under which such funding is provided. Wage levels were effectively not set by the Steering Group but by those external funding sources.

The wages of 'professional' staff were set by the operation of the local Council's job evaluation system. The wages of MSC-funded workers were set by the rules of the Community Programme. Budgets had to be produced and accounts inspected as a condition of continuing funding, and so limits were placed on how funds might be spent. Above all the plans for the redevelopment of the site were blocked by the limits on funding and new plans prepared based on possible funding assumed to be available. Moreover the issue of funding was reduced to one of a 'factual' question of how much was available, and the political questions about the needs and aspirations of the local community were side-stepped. The first plans for redevelopment were aborted when the tenders exceeded the funding allocated, and the Consultant brought in to undertake a financial appraisal of the Project. Despite requests that this be widened to include an economic and social appraisal, the Department of the Environment made no change to the terms of reference.

In the ensuing discussions the Community Council were repeatedly assured that the government was willing to support the Project provided that plans could be developed within the finances that were 'realistically' available. When pressed on how much was available no indication was given that there might be any more than what had been proposed earlier, that is, just over the £2 million jointly from Greater London Council (GLC) and Brent Council's Urban Programme allocation for the next three years. No assessment was made of the benefits, both socially and economically, which might accrue from a larger Project with a higher capital cost. The funding was thus treated as a cost, not as an investment. The uncertainty inherent in any patron–client relationship was a key feature of the Project from early 1984, as the Steering Group increasingly discovered that the only way that the Project could continue was by acceding to the demands placed on them to reduce the level of funding required. In order to do this, major changes to the scheme design were necessary.

Various attempts to make marginal changes and to retain the main proposals, in line with the aspirations of the Community Council, came to naught. In 'attempting the impossible' the Steering Group engaged in unproductive activities and held meetings that became more and more dominated by discussion of fine detail, away from the broader vision that had been the Project's foundation. Delay and uncertainty about the Project's future were carried through into the on-going work of the Project. With no agreement on the redevelopment the work plans for MSC-funded workers became short-term in nature, continually adjusted as delay increased. Day-to-day management became more a case of 'getting through' rather than based on clear plans for the work to be undertaken. Relationships became more and more strained, and manager–worker interaction alternated between attempts to impose control and resigned avoidance.

Bureaucratization

As the Project developed there was an increasing trend towards bureaucratic modes of organization. Each source of state funding carried with it a system for applying for such funding. In the case of the EEC and the MSC, both used as sources of wage funding, the systems were very complex and differed from each other in the manner in which expenditure was to be calculated and reported. Application for such funding was therefore effectively made by those who were able to understand the system and able to prepare the application. Moreover applications had to be made well in advance of the planned use of the funding. So a staffing structure was set up based on formal and abstract proposals which were likely to meet the funders' criteria rather than the emerging needs and plans of the Steering Group.

This led to a continuing process of dealing with problems caused by having a structure which did not really suit the needs as felt by those directly involved with Project. Thus the assistants to the Project Coordinator were required to act as supervisors of work groups because that was the 'only', as perceived, source of wages for them. Each of the 'professional' workers was required to take on a heavier work-load because their assistants were involved in other areas and, because *those* posts were funded no other funding could be obtained to pay others. So work overload was structured into the Project. However this was not recognized as such but, as for example stated in the report to the Policy and Resources Committee, 'inadequate servicing' by professional staff.

The relationship with Brent Council, inherently a patron–client one, developed into a situation where the Project was required to meet the administrative requirements of the Council. At first the Policy Coordinator acted very much as an advisor to the Steering Group, providing them with relevant information and advice. As the crisis emerged the Steering Group was increasingly required to provide the Policy Coordinator, and through him to the Council, with information *for* the Council. When the Consultant was ready to provide his initial assessment he met with Brent Council's Chief Executive, Policy Coordinator and the two councillors *before* reporting to the Steering Group. Later, when proposing that a Project Manager be appointed, the Policy Coordinator gave as a main reason that 'if the Project had a Manager with this overall responsibility then I would try and have regular meetings with that person instead of using the BoD meetings to *check up on particular points of concern to me*' (emphasis added).

The new management structure was to be established to deal with administrative problems that caused difficulties for the Council, and hierarchical structure was assumed to be the appropriate model to adopt. The origins of this tendency to bureaucratic organization may be seen in the way in which the Project was set up. The Project was not a community employment project in the usual pattern, a job creation project with clear and limited areas of activity. The Bus Depot Project was a project to set up a community project. The original vision of a centre in which the community would have access to social, welfare, educational, leisure and sports facilities, and assistance and premises for small business enterprises was effectively translated into one of having a project to produce these. The tasks of the Project were more and more concerned with dealings with public authorities, and less and less with the needs of the community. This diverted the Community Council membership from developing the action necessary to deal with the problems of the area in terms of the interests of the local community. The activities of the Community Council outside of the Project did not develop to any significant degree and much that was started soon faded. Instead the Community Council mainly concerned itself with the internal management and administration of the Project.

Ideology, Community and Project Management

Much of the development of the Project may be understood in terms of the way in which ideological hegemony is maintained by the state. Most commentators on the riots and their causes agree that there was not a coherent ideology underpinning the violent outburst: they were mainly spontaneous in character, in response to identifiable incidents within a context of disadvantage (Scarman 1982, Kettle and Hodges 1982). The Community Council's beginnings also were not based in any coherent ideology. The local Council and other government agencies gave credence to the Community Council's emerging role as the recognized 'voice' of the youth of the area; but the recognition of the Community Council by the authorities was very much on the basis of the few activists speaking 'for' the local youth. The four founders of the Community Council were to become members of the original Steering Group to examine the possibilities for the bus depot site. The report stated that

> The (community) council is a body of ten people, seven of whom were elected at an open meeting, the remainder being invited to serve by the elected members ... The council has no written constitution, no written rules, no formal system of membership and has not adopted many of the bureaucratic forms of administration which prevail in Britain today.

However, the Community Council became very much involved with bureaucratic forms of administration at an early stage. From involvement with the Youth and Community Service the Community Council moved straight on to involvement with a body set up by Brent Council, working with local councillors, assisted by the Policy Coordinator, in order to produce a report which would be presented to the local Council and other state agencies. The effect of this was that the Community Council did not have the opportunity to develop any coherent ideology of its own. There was no period of struggle against the authorities during which a framework of analysis of why the economic and social conditions were such that they were.

Instead the existing conditions were taken for granted; attention was placed on what changes could take place, but within the framework of understanding and interpretation of the authorities. So the use of the MSC as a source of funding was never really questioned; the involvement of local councillors as members of the Steering Group never challenged; the introduction of administrative procedures for obtaining funding never subjected to critical discussion.

In short, a group that may be described certainly as 'inexperienced', perhaps even 'naive', in terms of the political and administrative processes of the state was diverted from community action to managing a project whose agenda for direction and control was set by the state, not the Community Council. The *image* of community control was, however, maintained and reinforced by the management structure set up. In attempting to control the Project the Community Council placed considerable emphasis on details of decisions and actions at a tactical level. Steering Group meetings were held every week and items dealt were mainly of a short-term nature. Reporting back on details was the norm and meetings regularly overran. But aspects of policy and longer term thinking and planning were subject to limited discussion, and what discussion that did take place was mainly in response to agendas set by external authorities. The critical role of central coordination, ensuring that the Project was being strategically controlled in line with the interests and needs of the community, was not recognized. Even the discontent surrounding the Project Coordinator's performance centred on his poor administrative record, not the fundamental problems of the lack of the ability to respond adequately to the external pressures.

When the broader issues of policy and longer term strategic planning were raised by the Policy Coordinator, the wider economic, social and political issues were effectively removed from the potential agenda. In achieving this the possibilities of such issues being raised had been restricted. Crucial to this was the suggestion that the Steering Group had been 'badly advised' and 'poorly serviced' by the 'professional staff'. By implication, this suggested the Steering Group was in itself able to manage the Project *if it took the advice of the public authorities funding it*. The meeting between the Steering Group and the Consultant was closed to the 'professional' staff, and the main points raised were told to them after that meeting. The meeting with the Department of the Environment did not include any of the 'professionals'. The introduction of the new management structure was presented as superior to the previous structure, but without discussion of the purposes it was intended to serve. No explanation was provided as to how the functions to be performed by the Project Manager differed from those that the Project Coordinator was supposed to perform. The previous structure was described as 'cooperative', with no discussion of how such description fitted the nature of the relationships between the 'professional' staff, nor how the problems of the Project were related to the structure.

As the performance of the individual who held the post of Project Coordinator had been the subject of complaint and disciplinary action, it might very well be argued that there was no evidence at all on whether or not the structure was in principle faulty. But the needs of the state agencies involved were to have clear lines of contact with the Project through which control might be exercised. Thus a group which had spontaneously adopted, and been accepted as having, a leadership role concerned with community action was incorporated into state mechanisms before it could develop its own ideology. The large scale of the 'prize', the bus depot, led to an emphasis on obtaining large scale funding at a time that the state was taking steps to ensure such funding was used to harness any community initiatives to its own ends, that is to ensure that they helped to ensure the continuation of the capitalist nature of the state.

The logic of bureaucratic form of administration came with the involvement with the state bureaucracy, so that problems and opportunities were defined in terms of the needs and interests of the state. In effect a nascent community initiative was transformed into a state enterprise whilst giving the appearance of being under community control.

Chapter 10
Who Managed the Charlton Training Centre?

We noted earlier that the immediate context in which the Charlton Training Centre was established had a number of significant features. The effects of restructuring in capitalist enterprise had resulted in high unemployment and major changes in the nature of patterns of employment. The state intervened in the working of the labour market, in particular through the Manpower Services Commission (MSC) to enhance such changes. In London, the Labour-controlled Greater London Council (GLC) engaged in radical policies intended to promote the interests of Labour, and established the Greater London Training Board to develop training policies and activities. Moreover the GLC was committed to policies of promoting the involvement of ordinary people, through trade unions and community groups, in the various strategies being developed. The proposals to establish a training centre under community direction and control fitted well with such policies. We shall examine the major influences exerted by the state by relating in turn the effects of various actions by the MSC, the GLC and the Greenwich Council on the development of the Centre. The history of the Charlton Training Centre covers a period of just over four years, starting with the campaign to retain the MSC SkillCentre in April 1982 and culminating in the closure of the Centre in September 1986. Before we begin to examine the major influences, there follows a brief chronology of events.

The Origins of the Charlton Training Centre

The beginnings of the project lay with the Manpower Services Commission (MSC) issuing, early in April 1982, a consultative document on proposal to close the Charlton SkillCentre, as it was deemed to be surplus to requirements following the opening of a new, and larger SkillCentre in a neighbouring borough. Greenwich Employment Resource Unit (GERU) organized a campaign to oppose the closure. However, by 25 May 1982, the MSC announced its decision to go ahead with closure, to be phased over 12 months.

At the start of July 1982, the Greater London Council (GLC) called a meeting of interested bodies and groups to 'open up discussion amongst those agencies who had expressed concern over the proposed closure of the Charlton SkillCentre'. Following that meeting, between July and December 1982, preliminary development work on proposals was undertaken. Various options were explored by Greenwich Council, Greater London Manpower Board (GLMB), Inner London Education Authority (ILEA) as well as the 'Consortium' of local groups. The options were very soon displaced by the decision to go ahead on the basis of re-opening the SkillCentre after MSC closed it. To assist in taking this work further, the GLMB agreed to fund temporary development worker, and in December an appointment to the position was made. The development worker was formally employed by GERU on behalf of Consortium. A Sub-Group was set up to examine management structure of the Consortium/Centre.

In February 1983, an application was submitted for funding by the European Social Fund (ESF) for women-only training, the GLC to provide matching funding as required by that funding scheme. An application was made to the GLC to fund temporary outreach workers to develop links with local communities. In March, the GLC arranged to purchase equipment at the SkillCentre, from MSC, when it closed. However, the GLC Legal Department blocked the application for outreach workers.

In April, a draft staffing schedule was drawn up, and funding applications submitted to the GLMB. Discussions were held with MSC regarding setting up a managing agency under the newly instituted Youth Training Scheme (YTS). However, Greenwich Council indicated that it may possibly boycott of the YTS, in line with many other Labour-held local authorities. Concern was also expressed in the Consortium, in May, regarding the size of the proposed YTS scheme (120 places) and the consequent likely restriction on adult training at the Centre. The Development Worker reported that no MSC funding was available for adult training and expressed doubts on possibility of MSC agreement to mixed youth and adult training. Draft advertisements for jobs were agreed in May, and funding for advertisements was sought (under Chair's action) from the GLMB, now re-named the 'Greater London Training Board' (GLTB).

The application to the ESF had to be *resubmitted* in June because of queries over certain sections. A draft constitution for the Consortium as company limited by guarantee was discussed, and arrangements for recruitment agreed. Greenwich Council Industry and Employment Committee agreed funds for advertisements, and the Popular Planning Unit of the GLC stated that funding was available for consultancy. Further doubts were expressed, in July, over the use of the MSC's Youth Training Scheme. During July, agreement was reached to place advertisements for staff. A business plan was drawn up. In September an application was lodged for funding from MSC under the Youth Training Scheme.

The Greater London Training Board (GLTB) now reported doubts over the powers available to the GLC to provide funding to Consortium. It was agreed to await discussions by the GLTB on its funding powers. The GLTB also expressed concern that it seemed that courses were to be run on traditional MSC Training Opportunities Programme (TOPS) basis. It was agreed that sub groups would look at each course. The business plan was not submitted to the GLTB because of funding difficulties; the purchase of the SkillCentre was postponed, as was the recruitment of consultants.

One possible source of funding being explored by the GLTB was the use of a provision under local government legislation, whereby a local authority had the power to enter into arrangement with the Manpower Services Commission under any provision of the Employment and Training Act 1973. However, the GLTB's submission was blocked by MSC. Another source of possible funding that GLTB officers explored was that provided under section 137 of the Local Government Act 1972. This permitted expenditure to the level of the product of twopence in the pound from revenue raised from rates on households and businesses, for activities not explicitly provided for in other local government statutes. However, by this time, all the GLC's section 137 money had been spent or was already allocated.

In October 1983, GLTB officers expresses concern over diminished links between the Consortium and Greater London Enterprise Board (GLEB), and over the fact that package was 'too YTS and TOPS orientated and not as innovative as the GLC would have wished'. The Consortium reaffirmed its underlying philosophy of creating a new centre. GLTB officers asked to explore all avenues of funding. The Development Worker pointed out that funding of his post expired at the end of December, and stated that job had changed so that he now did a lot of administrative work. A meeting was arranged between GLC members and Consortium; however, this was postponed.

The following month, the GLTB expressed its commitment to Consortium. The MSC stated that the YTS places that had been reserved could no longer be held; this meant that all funding now had to be sought from the GLTB. The GLTB agreed a funding application in December, and advertisements for permanent staff were placed. In January 1984 the GLTB agreed funding for the temporary support workers and associated costs. A revised application to ESF was submitted.

In February 1984, shortlisting was undertaken of applicants for the posts of Trainee Support Coordinator[1] and for Training Coordinator. It was judged that there was no suitable applicant for Administrative/Finance Coordinator post. The site for the crèche was considered to a problem, as it was considered inappropriate for this to be housed at the Centre whilst building work was being carried out. Problems were also experienced in obtaining payments from the GLC.

At this time agreement was made that, as a matter of policy, the outcome of the recruitment process for coordinator positions should not be the appointment of two white workers. Also in February, the Consortium held a 'study day' to deal with issues about development plan, constitution and membership of Consortium, and employment terms and conditions. It was agreed that workers and trainees should have representation on Consortium.

Last minute problems were experienced with the lease, in March 1984: objections by the Head Lessees resulted in the Consortium accepting a condition in the lease agreement by which it waived automatic rights to remain as tenants at the expiry of the lease in 1987. The condition had to be heard in court so occupancy of the site was delayed. Greenwich Council withdrew its administrative support

---

1 That is, the Coordinator *for* Trainee Support.

for the Consortium, ostensibly because of pressures on services from within the Council.

The first permanent worker, the Trainee Support Coordinator, started in April, but there was a delay in start date for Training Programme Coordinator. The Consortium finally gained occupancy of Centre. At this time, the GERU member of Consortium expressed concerns about Consortium's employment practices being inconsistent with its espoused philosophy. The Consortium agreed with the views expressed.

Problems with the GLC Finance Department continued. In May, the GLTB received a progress report and granted funding for three month period ending June. The first courses were targeted to start in August or September. Discussions were held on the management structure of Consortium and Centre: it was agreed to set up sub-groups to be 'responsible for intensive thinking and support around aspects of the Centre's work'.

A further progress report was made to the GLTB in June and further funding to the end of August was agreed. However, payments from the GLC were still outstanding and the workers objected to advertisements being placed for further workers whilst the pay of current workers had not been released by the GLC. At this stage, a disagreement arose between the Consortium Development Worker and a member of Consortium, and in July, the staff wrote to Consortium about grievances. The start of courses was delayed until November because of delays in the building work. The nursery was also delayed, and it was agreed to operate an interim child-minding scheme. However, the GLTB did approve funding of £2.5 million, to run until December 1985.

The Trainee Support Coordinator had presented a discussion paper on management structure, in June, and this was discussed in July. It was agreed that the Centre Coordinating Team have responsibility for 'smooth and efficient running of the Centre'. A reference to the Women's Unit as 'semi-autonomous' was deleted and it was stated that the Unit was 'an integral part of the Centre'. In August, a structure was proposed by the GERU representative, setting out responsibilities of Coordinators and sub-committees; this was agreed.

In October, the Consortium and staff met to discuss the grievances raised by staff. The Consortium presented a statement regarding organization of the Centre. Under this, the Board would decide policy on basis of detailed policy options presented by relevant Centre Coordinators; each member of staff would be responsible to a designated Centre Coordinator. The Finance and Administrative Sub-Committee and Staff Sub-Committee would continue, and a new Centre Coordinating Sub-Committee was set up; other sub-committees were to be wound up.

The first course eventually started in November 1984. Appointments were made to the positions of Administrative/Finance Coordinator and Development Worker. The Chair of the GLTB addressed the Consortium, raising various concerns about the Centre that had been referred to him. In January 1985, staff expressed concern about mounting problems at the Centre and asked for five-day shut down to deal with them. This was refused by the Chair. A progress report was made to the GLTB, which asked for a further report in March. By this stage, the Government had embarked on its plans to gain Parliamentary approval to abolish the GLC. The next month, the GLTB representative reported that such Parliamentary action meant that funding was still under threat and emphasized the need for monitoring reports.

Staff contracts were finally agreed in February 1985. Trainees in the Women's Unit petitioned the Chair on matters of concern to them, and the Board discussed this after asking the Women's Unit Coordinator to leave the meeting, to which she objected. The Trainee Support Coordinator stated that she was unclear about relationship between herself and the Women's Unit Coordinator and asked to be relieved of responsibility for the Women's Unit. The Chair refused this request. The following month the Women's Unit Coordinator resigned. A meeting was held regarding the Women's Unit, and the GERU representative asked to be relieved of responsibilities for Unit. It was agreed that workers of Women's Unit be given opportunity to resolve internal problems.

Also in March, the Board decided that when matters of confidentiality relating to Centre workers or trainees were being discussed, Trainees' Representatives would be asked to withdraw from the meeting. The GLTB approved the Centre's budget for period to December 1985, and approved continuing grant. The Centre workers and Consortium produced a Progress Report for submission to the GLTB

In April 1985, the Board held a special meeting to review the management structure, at which the Centre Coordinators presented proposals. It was agreed that each worker would be accountable to a specific Centre Coordinator. However, the relationship of the Women's Unit within the agreed structure was not decided, and was to await outcome of the discussions within that Unit.

Two months later, in June, the Trainee Support Coordinator resigned. The Chair now appointed a personnel officer without reference to the Board and without advertising the post. The Chair then went on a trip to USA, and whilst they were away the black workers' group raised objections to the appointment to the personnel officer. The Board decided that the post be held in suspension but that the appointee be kept on as an employee. It was also agreed that proposals on disciplinary and grievance procedures were to be circulated, for discussion at next meeting, in July; however, this was not done in time for that meeting.

At the July meeting, the Chair defended the decision to appoint a personnel officer. The Board meeting debated the issue but the voting was tied. The Chair decided that it would be improper to use the casting vote, and it was agreed that the matter be referred to next meeting. However that next meeting failed to discuss the matter, which was not formally raised again until September. Meanwhile, the person appointed continued to work as the Personnel Officer.

In August the Board dismissed the Administrative/Finance Coordinator. Temporary appointments were made of a member of the Education Unit as Acting Administration Coordinator, and of the Treasurer as Finance Coordinator. At the September meeting of the Board, workers' representatives again raised the issue of the appointment of the Personnel Officer; however, the meeting became inquorate before a decision was reached.

In October, a dispute arose between the Acting Administration Coordinator and one of the Outreach Workers over the use, by that Outreach Worker, of the outreach office as a waiting area for interview candidates. The Acting Administration Coordinator suspended the Outreach Worker. The action by the Acting Administration Coordinator was confirmed by a meeting of the officers of the Board. The staff questioned the procedures adopted and pointed out that, although no procedures have in fact been agreed, the procedure actually adopted was in breach of both the proposed procedures and the guidelines set by the Advisory, Conciliation and Arbitration Service (ACAS). In November, the GLTB Chair called for a meeting with the Consortium, workers and trainees to discuss how problems of Centre might be overcome.

The Consortium Annual General Meeting was called in December, but this was initially inquorate. At the second attempt to call the Annual General Meeting, the posts of Secretary and Treasurer were taken on by new members; the existing Chair was re-elected. An application to the GLTB for continued funding was passed to the GLC Policy and Resources Committee which agreed in principle.

At the beginning of January 1986 there was still uncertainty about state of the Centre's finances whilst awaiting for the GLTB's decision. As a result, one of signatories to the Consortium cheques refused to sign for any items except childcare expenses and trainees allowances. The GLTB finally agreed, during the month, to fund the Centre until September subject to further review in March. However, such funding was subject to the consent of Secretary of State, as it extended beyond the abolition of the GLC at the end of March, as now agreed by Parliament. That consent was forthcoming in March, and at its last meeting the GLTB agreed the forward funding.

Following abolition of the Greater London Council, on 31 March 1986, the newly formed London Boroughs Grants Committee took over responsibility for funding for voluntary sector organizations operating across a number of London boroughs. During the period from April to August 1986, discussions were held with officers and members of that Committee and of Greenwich Council regarding continuation of funding. Officers proposed restructuring of the Centre's operation and management. However, in September 1986, the funding application to London Boroughs Grants Committee was rejected, and in the absence of continuing funding the Centre closed.

Analysis

The Manpower Services Commission: A Diversionary Legacy?

A report to the March 1985 meeting of the Greater London Training Board in describing the way that staff of the Centre have had to develop 'from scratch' detailed policies (on training process, equal opportunities, and other matters) stated that 'the inheritance from the MSC, other than a building and some equipment, was nil'. This statement, however, can be seen to be a rather limited view of the influence the MSC on the development of the Centre. In reality, the MSC, in a variety of ways, exerted considerable influence. The MSC was a major instigator of the Consortium in the sense that, by closing the Charlton SkillCentre, it provided a number of individuals and groups with a focus for engaging in collective action initially to oppose the closure and later to create an 'alternative'. By vacating an existing site specifically adapted for training activities it provided the Consortium and the various interested parties with a physical base within which such an alternative could be set up. Moreover during the development stages the MSC influenced the direction such development took by virtue of the dominant role it has over the ideology of training. Finally, insofar as the MSC was an agency of a central Government intent on limiting the role and functions of local authorities, especially local authorities committed to policies opposed to those of central Government, it had considerably influenced the actions taken by the Greater London Training Board.

The Closure of the SkillCentre

The closure of the Charlton SkillCentre was proposed by MSC as part of a programme of rationalization of the SkillCentre network. A new SkillCentre was shortly to be opened at Deptford which would be underutilized. By transferring all classes from Charlton the new SkillCentre would be able to operate more fully with no loss of training places available. In answers to written questions put by the local MP, the Minister responsible claimed that the average costs of training would be considerably less following the transfer than the average costs then current at Charlton. The proposals were criticized by the Greater London Manpower Board (forerunner of the Greater London Training Board). The report argued that any short term savings would be at the expense of weakening the potential for expanding skills training provision in the event of any upturn in the economy. Estimates of skills mismatch in the London labour market could account for over 100,000 of the unemployed by 1984. The report drew attention to the ideas behind the MSC's 'New Training Initiative' published in late 1981 which were yet to be put into effect, and to the Government's stated policy of improving the skills of the workforce to take advantage of technological developments.

The closure of Charlton was, the report stated, the loss of a fifth SkillCentre unit in London. However, whilst these 'rational' debates were taking place, more comprehensive critical analyses of MSC policies and practices were being developed. Criticism of the discriminatory practices of the SkillCentres was being made. Various women's groups formed the Women into SkillCentres Campaign which drew attention to the fact that only 3 per cent of SkillCentre trainees were women, and produced a set of demands to the MSC in order to bring about a change in this situation. Among the demands were: nurseries at all SkillCentres, childcare allowances, introductory courses for women only, flexible training hours and women instructors. The Campaign also demanded a clear MSC policy opposing sexual harassment.

At the 'Training in Crisis' Conference organized by the GLTB in May 1984 a workshop session on SkillCentres and Adult Training reiterated these demands and also stated that

> Adult Training in general and SkillCentres in particular grossly under-represent black people, both as trainees and especially as staff. A race conscious policy on recruitment and positive action is urgently needed to redress the balance. (Greater London Training Board 1984b)

The workshop did, however, recognize the need to defend the SkillCentre network against cuts and closures and argued for a mixture of public sector provision and community and local initiatives.

A more comprehensive analysis of MSC policy on training was also being developed at this stage. This analysis placed the actions of central Government on training in the wider economic, social and political policies being implemented by Government. We noted earlier in examining the context of the establishment of the Charlton Training Centre some of the actions taken. The chronology of action taken by the Government, both directly and through the agency of the MSC, can be summarized as follows:

- 1979: ten per cent cuts in funds for Industrial Training Boards and Colleges of Further Education offering Training Opportunities courses (primarily clerical training for women).
- 1981: 'Sector by sector' review of industrial training arrangements carried out by MSC; despite recommendations that no ITBs should be abolished, Government announced plans to close 16 out of 24 and transfer of funding for the remainder from MSC to employer levies.
- 1982: ITBs closure programme was carried through; the SkillCentre closure programme started.
- 1983: the SkillCentre Training Agency was established to operate at 'arms length' from the Commission subject to meeting commercial objectives in competition with other providers of training. The Community Enterprise Programme (job creation scheme for long term unemployed) replaced by the Community Programme with lower rates of pay and restrictions on provision of training. Youth Training Scheme started.
- 1984: MSC announced intention to lose 29 SkillCentres.

The conclusion reached by the critical analysis of these events is encapsulated by a statement from the workshop at the Training in Crisis Conference, that 'the aims and practices of government training, particularly that of the MSC, is to divide the workforce into an educated élite and an un- or semi-skilled mass (as well as to reproduce race and sex divisions)' (Greater London Training Board 1984b).

Such an analysis clearly formed a key part of the initiative taken by the worker at Greenwich Employment Resource Unit in organising opposition to the proposed closure of Charlton. GERU's response to the 'consultation' carried out by MSC included a document, 'Training: The Potential for Expansion and Development', setting out ways in which the training needs of local people could be met, linked to job creation particularly through self employment and cooperative enterprise, and following a policy of redressing the discrimination against women and ethnic minorities. The GERU worker was an active member in the Women into Manual Trades movement. She argued for the views expressed in the response to the MSC at meetings of the 'informal Consortium' set up following a meeting called by the GLMB on 1 July 1982, and argued that the Consortium should not tie itself necessarily to keeping open, or reopening, the Charlton SkillCentre; rather, it should be prepared to consider provision spread over a number of smaller sites.

However by the beginning of November 1982 the Consortium had 'broadly agreed that one major training centre, probably on the Charlton site, was needed'. This point of view was one which was argued for by the Greenwich Council representative in a paper presented to the September meeting of the Consortium and appears to be favoured by GLMB officers. After the November meeting all discussions and development work were based on the concept of taking over the Charlton site. Throughout the discussions during the meetings in 1982 there was a sense of urgency, often reported in minutes and documents, with the MSC closure timetable and GLC and Greenwich Council funding timescales forming the timescale to which the Consortium had to work. Clearly then, the development of proposals and plans for alternative provision, although arising within a context of criticism of MSC policies and practices and of demands for alternatives, was shaped by the MSC's decision to close Charlton and the arrangements for so doing. The Consortium's proposals arose primarily as a response to the closure rather than as a separate development.

MSC Influence on Training

We noted earlier that the MSC had played a significant part in the state's labour market policies and strategies. One key element in this had been the concentration of influence over training on the MSC's sphere of activity. Other influences had been successively reduced in influence so that the MSC held almost unchallenged sway over the very perceptions about what training is for and how it should be conducted. This trend is of particular relevance to the Charlton Training Centre. The 1973 Employment and Training Act by which the MSC was set up was described by the GLTB as a 'compromise' between those in the Conservative administration who wanted the ITBs to be abolished and opponents which included trades unions, employers, educationalists, and professional trainers.

> In the event, the Act compromised. ITBs were retained but their key power to raise levies from industry was weakened and they became less effective as the economic climate deteriorated during the 1970s. The Act gave the trade union movement (then enjoying considerable influence over government policy in the wake of successful industrial struggles in the early 1970s) something that it had been demanding since the end of the Second World War – a central statutory labour market organization – the MSC. (Greater London Training Board 1983: 5)

The MSC was intended to undertake a coordinating role with regard to the public employment and training services. However with the increase in unemployment especially among school leavers the MSC was increasingly called upon to develop special measures as 'constructive alternatives' to unemployment. By 1978-1979, over 160,000 school leavers were on Youth Opportunities Programme schemes and a total of £76 million pounds was being spend by the MSC on schemes for long term adult unemployed. Much criticism of the MSC activities focused on accusations of 'massaging the unemployment figures' and of attempting merely to contain the social and political effects of unemployment rather than directly to tackle unemployment. The GLTB's analysis goes further, noting that Conservative governments have traditionally sought to use labour market policy to solve economic difficulties:

> At the same time as strengthening the state there is also a conscious effort to reduce the training policy functions which the state directly carries out and to strengthen the hand of employers in training policy matters. The Government's position in training policy is fully consistent with its anti-trade union attitude in labour market policy generally. (Greater London Training Board 1983: 5)

One of the major effects of the cuts in funding and then the transfer of funding of ITBs from MSC to employer levies was the reduction of research and development carried out by ITBs. Much of this research and development was carried out cooperatively between ITBs through Inter-Board Study Groups in particular areas such as supervisory training, industrial relations training, management development, women in management. Many ITBs had developed comprehensive standards based modular training schemes at craft and technician level and others were well advanced in this work. In the Furniture and Timber ITB final agreement with the trades unions to introduce such a scheme as a development of the apprenticeship scheme was conditional on the retention of that ITB, which did not happen.

At the same time as the pioneering work and the research and development expertise in ITBs was being lost, the MSC increasingly intervened in such work in a highly directive way. By 1983 all MSC funded research was required to have a direct link with the work of the MSC as stated in the objectives of the New Training Initiative, as determined by the MSC. In particular the first objective, the introduction of standards based skills training, provided the basis for major funding of research into the development of approaches using the concept of 'occupational training families'.

Earlier, in 1980, the Central Policy Review Staff had argued that the concepts of 'skill' and 'apprenticeship' were anachronisms with no place in a modern economy (Central Policy Review Staff 1980). Instead, argued the report, these concepts should be replaced by the more appropriate 'concept' of competence as determined by employers in relation to the immediate requirements for task performance. These ideas played a considerable role in the development of YTS and were used by the MSC as the basis for deciding on approvals for applications for individual schemes under YTS funding. Training for the unemployed was mainly funded by MSC through the Training Opportunities Scheme (TOPS). Industrial Training Boards were restricted by the terms of the Statutory Instruments under which they were set up and could only use special funding (not operational funding provided by MSC or employer levies) to undertake training for the unemployed, who by definition were not employed within the industries served by the ITBs. Most TOPS training was carried out within SkillCentres, the network of centres operated by the MSC.

The MSC was then the main agency for training for unemployed adults and the pattern of provision was the main 'bench mark' against which alternative provision could be compared. Such comprehensive grip over the ideology of training provision can be seen to have influenced the developments of the Charlton Training Centre. During the first few months of the Consortium's development discussions were being held with the MSC, which had a representative on the Consortium, regarding the possibility of setting up a training workshop under YTS funding. The Greenwich Council representative was separately negotiating with MSC about the possibility of the Borough Council taking over the Centre to set up such a training workshop. Only in May 1983 is any mention recorded in Consortium minutes regarding doubts about the YTS, and even that is in regard to the size of the scheme being considered compared with the size of adult training provision. MSC were at this stage provisionally allocating a number of YTS places for the Centre. In July the representative from the GLTB expressed concerns about the standards of training which would be provided indicating that if the GLC were to provide funding the standards would have to be higher than those under the YTS. Even then a submission was made for MSC funding in September 1983. The GLTB representative repeated concerns over YTS and also the fact that adult course proposals appeared to be based on TOPS provision.

Delays in obtaining preparing a comprehensive funding application resulted in the MSC re-allocating the provisionally reserved YTS places and the Consortium was left with only the GLC (and EEC) as the funding source. But even as late as March 1985 the Consortium was taking TOPS as the base for comparison, stating in a progress report to the GLTB that it was 'aiming to provide courses which are comparable with TOPS courses for skill level'.

Thus throughout the development of the Centre the MSC-defined notions of training have played a considerable role, despite the concerns expressed by the major funder which itself has developed a comprehensive, coherently critical analysis of the MSC and its programmes.

Restrictions on GLTB Funding

Finally, the MSC played a restricting role on the GLTB's ability to support the Consortium. This arose through the problems over the use of 'Section 45' funds. This refers to section 45 of the Local Government (Miscellaneous Provisions) Act 1982 which empowers local authorities to enter into arrangements with the MSC under any provision of the Employment and Training Act 1973. The GLTB had aimed to use such funding under section 45 for the Charlton Training Centre but in September 1983 was informed by the MSC that this would be outside the powers in section 45 of the Act. The uncertainty about funding which preceded this statement by the MSC caused delay in a funding application being put to the GLTB.

This delay came at a critical stage for the Consortium's development work. An application had been made to the European Social Fund for women-only training and an overall application for GLTB funding had been prepared. Advertisements for staff had been placed in anticipation of receiving GLTB funding and completed application forms were being received. The delay not only caused disruption to development work but also seriously damaged the morale of members of the Consortium. The then Chair of the Consortium wrote to the Chair of GLTB expressing such concerns pointing out that

> at a time when increasing demands are being put on their [i.e. voluntary and community groups] over-stretched resources, such groups have been, and are still, giving the considerable time and energy necessary to formulate proposals which ensure that Charlton will live up to its philosophy in practice as well as theory.

The Chair went on to state that the Consortium was also concerned that 'further delays in considering proposals, and in failing to take decisions that will allow Charlton to start changing from a concept into a reality can only damage the goodwill, effort and momentum that has existed thus far'.

Thus the intervention of the MSC in blocking the anticipated use of section 45 funding took out of the hands of the Consortium considerable areas of management and seriously affected the Consortium's perceived ability to control the direction of the Centre's development.

The Greater London Council: More Than a Funder?

Part of the State The GLC had been significantly involved in the development of the Consortium throughout. Indeed the initial meeting was called by the GLC at the suggestion of GERU. The GLC was a member of the Consortium with an officer of the GLTB Support Unit acting as the nominee of the GLC and as a director.

The GLTB provided a grant to fund the Consortium's Development Worker who, the Consortium considered, was essential if the detailed work needed in developing proposals for funding was to be carried out. The GLTB agreed to provide 'matching' funds in support of the application to the European Social Fund for women-only training provision. The GLC provided a capital grant to help take over the lease of the site, to carry out building work, and to buy equipment from MSC. GLTB agreed to provide the funding to run the Centre and pay allowances for trainees.

In a very tangible sense, then, the GLC has made the reopening of Charlton Training Centre possible. However the GLC's influence over the direction taken, and the manner in which the work of the Consortium and the Centre was managed, was much more than that of 'funder'. As the 'Capital and Class' editorial collective stated, in writing an account of the first five months of the GLC's Economic Policy Group, 'we must keep reminding ourselves that GLC is part of the State, whatever its "relative autonomy"...' (Capital and Class Editorial Collective 1982: 132).

The ruling Labour Group on the GLC attempted to carry through its espoused socialist manifesto on which it won the 1981 GLC election, and to introduce a range of radical changes to the structure of GLC administration, but within the limits prescribed by law. Inevitably contradictions arose and these can be observed in the relationship of the GLC to the Consortium. In particular they may be observed in the administration of funding, and in the policy aims of the ruling Labour group.

Administration of GLC Funding The funding for Charlton was provided through the various formal and informal procedures adopted by the GLC. Despite the radical style of local government introduced by the ruling Labour Group, the GLC administrative machinery was highly bureaucratic in nature, in both the technical (Weberian) sense and the popular sense of the word. All funding by the two committees mainly involved with Charlton, the GLTB and the Industry and Employment Committee, required a committee paper describing the purpose for which the funding was sought. A concurrent report was also required from the Comptroller of Finance and also one from the Solicitor to the Council. The Solicitor reminded the committee of its duties:

> In deciding whether to exercise these powers the Board must act reasonably, which means that it must take into account relevant factors, disregard irrelevant ones and, having done so, not make a decision which no reasonable authority could have made. Also, the Board must have regard to the Council's fiduciary duty to London ratepayers to balance fairly the interests of the ratepayers against the interests of those expected to benefit from the expenditure. The Board should satisfy itself that the benefits to be derived from the expenditure warrant and are commensurate with it, and it will be financially prudent for the Council to incur the expenditure.

At a fairly critical stage in the Consortium's development the Finance Department blocked an application to the GLTB for extra workers to help develop the level of participation by various local groups and by the community as a whole. The GERU worker involved at this stage describes the process of applying for grants as

> being about producing a document by a specific closing date, written in a particular language and style, politically in line with the policies of the particular funder you are going to approach, and producing an acceptable budget which you can usually base on other people's to save you work. (Mantle 1985: 11)

The particular incident where the funding was blocked resulted in the outreach work planned not being undertaken as desired. This put more pressure on the members of the Consortium and on the one Development Worker who had originally been employed on only a four month contract. It also resulted in the meetings of the Consortium concentrating mainly on technical details of the developing proposals to the neglect of the policy and 'philosophical' issues inherent in a radical innovative community venture. The GERU member of the Consortium challenged the notion that the development of the Charlton Training Consortium was an example of popular planning. She did so not on the basis of criticism of the motives or intentions of those involved but because of the practical outcomes of the institutionalized practices of the parties involved.

> The Consortium was to be representative of different groups' needs. The emphasis was to be on the planning being done by the local community – this was what I and a few others wanted, and it was what the GLC, a potential funder, wanted. However, the process of getting the funding was to be no different; in fact it was going to be even more complicated than for most projects as a massive amount of money was needed, it also required support and cooperation of various funders and organisations, not just one or two. (Mantle 1985: 4)

The style adopted by the Consortium, frequent formal meetings dealing with highly technical matters with experienced 'professional' individuals undertaking the work involved in generating policy proposals, prevented it from encouraging and enabling individuals and groups in the community from participating in the development of the Centre. Insofar as the administration of the funding process played a formative role in this tendency the GLC significantly influenced the development of the Consortium and its organization and management.

GLC Policy Aims

We noted earlier that the GLC's clear set of radical policies, and the commitment to the participation of community groups in the implementation of those policies contained inherent tensions. There were tensions between various areas of policy commitment, and also between the determination to ensure implementation and the desire to gain community participation. This can be seen happening in the development of the Consortium. The decision to concentrate on developing plans based on re-opening the Charlton Centre, rather than plans based on dispersed sites, was taken very early on in the Consortium's development partly because GLC officers were making it clear that support would only be possible on this basis. The response by the GERU representative to this was that the Centre should provide training for local people only. However when the Centre had been taken over from MSC and detailed planning of courses was being done the GLC made it clear that training places should be advertised widely over London. This created considerable administrative problems for the Centre because of the high response rate. Moreover it cut across the work of the Outreach Unit of the Centre which was attempting to attract trainees from the various communities in the *local* area.

The strategy of the GLTB for training provision in London was of course developed in the context of national government policy and strategy mainly carried through by the MSC. The GLTB was in effect, espousedly developing a counter-strategy, and was publicly critical of MSC and government policies and actions. When the Consortium was developing its plans for training, GLTB officers expressed concern at the apparent relatively uncritical adoption of MSC standards (YTS and TOPS) as the basis for those plans. At a Consortium meeting in October 1983 the GLTB representative referred to the package of proposals being developed as 'too YTS and TOPS orientated and not as innovative as the GLC would have wished'.

In a letter responding to concerns expressed by Chair of the Employment and Industry Committee of Greenwich Council over problems with section 45 funding the Chair of GLTB stated that

> At a review meeting with GLC officers, called at my instigation, on 1 November in County Hall, we established the following points: 1. our strong political commitment to the project as an innovatory high quality training scheme was reaffirmed.

At about the same time other officers of the GLTB became involved to help with the details of the training proposals. As a condition of funding the Consortium was required to make periodic reports to the GLTB. The report in March 1985 was a detailed description of work undertaken by the various Departments of the Centre and caused considerable extra work for staff who were aware of the political implications of the report, that is, as a justification for continued funding. Thus while being clearly supportive of the aims of the Consortium the GLC has exerted considerable influence over the detailed planning and implementation of the plans, to ensure that policy conflicts are kept to a minimum.

Greenwich Council: A Community Base?

The influence of the Greenwich Council has been in two stages. Initially there was considerable support and at one stage the Employment Development Officer was chair of the Consortium. The Council provided the funding for the rental of the site, through Urban Aid, the urban regeneration funding scheme. However there were considerable tensions in this involvement which reflected the Council's own activities in the training field. In July 1984 the (second) Employment Development Officer was instructed, by the chair of the council committee to whom she reported, to cease her involvement with the Consortium. In October that committee decided not to take up the corporate directorship of the Consortium. Eventually, in early 1986, the Council became involved again. As the Consortium sought continued funding, and in the context of the reorganization of funding arrangements because of the abolition of GLC, the Council began to consider what form any support for the Centre should take. This took the form of seeking to change the nature of the centre so that it would take on a structure more in line with other training centres supported by the Council.

From Directiveness to Disengagement

The first phase of the Council's involvement may be characterized as moving from attempts to engage in considerable directiveness, followed by total disengagement in July 1984. The first Employment Development Officer (EDO) was an active member of the group which became the Consortium. His work with the Council involved establishing training and temporary work schemes. Many of the Consortium meetings were held at the Town Hall and administrative resources were made available through the EDO's office. However, it is clear that there were tensions between the proposals and actions of the EDO and those of the representative of GERU. The EDO prepared a paper for the Consortium, for discussion on 14 September 1982, in which he argued against the development of a number of dispersed training centres, saying that using the Charlton premises 'could turn out much cheaper than a range of smaller individual projects. This is especially the case if a sound package could be worked up with the MSC'.
This directly countered the view expressed by the GERU representative at the previous meeting, who reported to colleagues at GERU that she had argued that

> It might be more applicable to have training provision at various different sites, rather than just seek ways of taking over the Charlton site. Training should be more local, the catchment area for MSC skill centres are absurd, that is, massive.

She recalls that at the meeting on 14 September there was a

> Constant power battle between [the EDO] and the community groups. Had to continually argue that our proposals were practical and that there was a need for them. Very basic arguments over why we need childcare, why women don't do manual skills training and so on.

It also appears that the EDO was having separate discussions with the MSC about the possibility of using the site as a youth training centre. No minutes of the meeting were presented to the subsequent meeting. The next planned meeting was cancelled by the EDO who sent a letter out just four days before the date set. He set a date for a meeting in November stating that 'it is hoped that stemming from the next meeting, firm progress can be made in designing an innovative training scheme which can be funded through the MSC and possibly the GLC and LB Greenwich [Greenwich Council]'. The GERU worker recalled that, in her view, he was 'basically unable to comprehend that the GLC would fund the sort of proposals which were coming forward'. The GLC officer involved at this stage recalled that his main contribution was to keep reiterating that the GLC would be prepared to fund the proposed training centre. Other issues of contention that arose over the next few months included criticism of the EDO for not arranging a crèche at Consortium meetings, and of proposals by the EDO that a crèche at the proposed centre should be staffed by workers funded through the MSC's Community Programme, which would not provide for decent wages and conditions. By this time the Council was establishing a 'Greenwich Development Company' which acted as a Community Programme Agency and would establish a Youth Training Scheme.

Thus from the start Greenwich Council, through the actions of its Employment Development Officer, was attempting to pursue policies which emphasized speed and low cost, utilising the schemes of the MSC. The political issues on which the original campaigning had been established were regarded as secondary issues. As the possibility of MSC involvement waned, and the commitment of the GLC was strengthened, the influence of the Council declined. The active involvement of the Council ended in July 1984, after the Council member who was vice-chair of the Council Committee concerned left a Consortium meeting where he was informed that he was not entitled to vote as *he was not a Board member*. This came about when he tried to oppose a (successful) proposal that statements made orally to staff about holidays should be honoured. The meeting was being chaired at the time by the (replacement) representative from GERU.

In October 1984 the Council Committee discussed the invitation to become a corporate director of the Consortium. The Council's solicitor's comments included

> it is unwise for an individual, and even less so for a body corporate, to become director of a company with whose daily affairs he/she or it will not be concerned, or over which he/she or it cannot for any reason expect to exercise a reasonable degree of supervision.

The committee decided not to become a corporate director.

Re-engagement

Greenwich Council did not become actively involved with the Consortium until late in 1985. By this time the legislation to abolish the GLC had been passed and the future funding position was very uncertain. The Council agreed to take on supervisory responsibility for the forward funding provided by the GLC, but required a fee for this service. The discussions which then took place between the Council's officers and the officers of the London Boroughs Grants Scheme were clearly concerned with changing the nature of the organization and management of the Centre. The discussion document which they jointly prepared proposed that the funders should have a majority control on the Board, and that an overall general manager be appointed accountable to the Board. The report to the Grants Committee in September 1986 stated that the proposals put to the Consortium also included a 'complete restructuring and rationalization of the staffing of the Consortium ...[and] a review of salaries of the employees of the Centre which are typically higher than their equivalents in local authorities'.

Thus again the concerns revolve around issues of control by the local state, and reductions in the costs of the Centre, in particular the staffing costs. The final sanction which the state had in these circumstances, the withdrawal of funds, was applied, leaving the Council free rein to develop the plans for the Centre without 'interference' from those who claim to represent the community.

Conclusion

It is thus appropriate to examine the development of the organizational form and management approaches which emerged in the Charlton Training Centre in terms of the influences of the state, in the form of the MSC and the two Councils involved. Although the London Borough Grants Scheme regarded the problems which arose as 'internal' issues (lack of clarification of the respective roles of the Board and the staff), we can see how such problems arise in the wider societal processes which make themselves tangible in the actions of the state, centrally and locally. As Benson points out

> The production of organisational life is integrally connected to the social totality. The generative mechanisms, the contradictions, and the organising actions of the larger totality are involved in the production and reproduction of organisational practices. Dialectical analysis, then, must be concerned with how the developmental tendencies of the organization are guided by the tendencies of the totality. (Benson 1983: 337)

The tendency was similar to that we saw in the case of the Bus Garage Project: the transformation of an embryonic community organization which potentially challenged the structures which gave rise to the disadvantage and discrimination experienced by the community. The transformation was to that of an organizational form which reflected the form adopted by the state, thus ensuring that any potential alternative was undermined. Insofar as the organizational form adopted by the state was predominantly geared towards sustaining the interests of capital, to the disadvantage of labour, the net effect of the events at the Centre, as at the Bus Garage Project were contrary to the original espoused aims.

Chapter 11
Attempting Democratic Organization: Tensions and Contradictions

We have seen in the previous chapter how the various influences of local and central state agencies had significant effect in shaping the development of the Charlton Training Centre. The description of the organizational structure and management of the Centre as 'democratic', with full participation by the trainees, the workers, and the local communities is therefore seen to be simplistic and misleading. However, unlike the Bus Garage Project, there was from the early stages an explicit commitment to develop such a democratic structure. It is therefore important also to examine the steps taken by those involved in relation to such a declared commitment. From such an examination we should attempt to identify those aspects which indicate how and where problems arose. In this way the lessons may be learned and more widely spread. We shall firstly recount the relevant aspects of the history of the Centre up until August 1986, then analyse the processes at work which served to prevent the full realization of the espoused democratic structure.

Early Developments

The idea that Charlton should be organized and managed in some form of 'democratic', 'non-hierarchical', 'collective' manner existed from the very early stages. Certainly by November 1982 the application for funding for a development worker stated that the duties of that person included the incorporation into all plans developed by the Consortium the principle of community involvement and trainee and worker participation. This should not be unexpected given that the initial impetus, and a considerable amount of development ideas came from the Employment/Training Development Worker at Greenwich Employment Resource Unit (GERU), which acted mainly as a local *cooperative* development agency. It was also GERU which employed the Development Worker, because the Consortium was unincorporated and legally unable to employ staff. The GLC was prohibited from providing funding to an unincorporated group.

At the December 1982 meeting of the Consortium a sub-committee was set up to develop proposals for the organizational structure and management of the Centre. A discussion document was prepared by the Development Worker for the first meeting of the sub-committee, the document being titled 'How Can New Training Provision in Greenwich be Run on Democratic Principles? Cooperative

Management Structure at Charlton SkillCentre'. An impressive list of points were made including an examination of who had interest in the management structure, the constraints on decision making, the key decisions to be made and some outline proposals for the structures for decision making.

By late January 1983 the Sub-Committee had met, agreed that some adjustments be made to the discussion document, and decided that a solicitor 'with the appropriate expertise on legal structures' should be engaged to draw up the relevant legal documents. By the beginning of March 1983 a proposed constitution for the Consortium had been drawn up by a sub-group consisting of the Development Worker, a Greenwich Council officer who was also then chair of the Consortium, and an officer from Greenwich Council for Racial Equality (GCRE). Amongst the 'Objects' of the Consortium was included 'to agree a democratic management structure for the Charlton Skills Centre which ensures full participation by all those concerned with the operation of the Centre'.

Developing the Formal Constitution

The need for a formal constitution became urgent when, early in 1983, it was proposed that the Consortium employ ten part-time outreach workers who would work with the local community presenting the Consortium's ideas and plans and seeking views and ideas from people locally. The GLC officer reported to the Consortium's meeting on 17 March 1983 that 'GLC Legal Department had refused to approve funding for Sessional Outreach Workers because it felt that the Consortium's Constitution was inadequate'.

At the next meeting another GLC officer suggested that if GLC Legal Department refused to ratify the Constitution a company could be purchased 'off the shelf'. However a solicitor was appointed shortly afterwards and by June had devised a proposed Memorandum and Articles of Association for a company limited by guarantee to be set up by the Consortium. After some amendments were made to the draft Memorandum and Articles the company was set up as a 'corporate shell', having three named subscribers, the GERU worker, the Development Worker and the representative from the Greenwich Afro-Caribbean Association. As this work was being undertaken the Consortium had also been developing its plans for the Centre's activities and the staffing arrangements deemed appropriate. At the meeting on 14 April 1983 the Consortium agreed that

> a Coordinating Management Team of four people be appointed. This would consist of one Training post, one Support Services post, the other two posts not being so well defined. However, these would require skills in finance and/or administration. When advertising for these posts it would be stressed that each coordinator would be expected to work as part of a team.

The agreed salary level for these posts was the bottom of local authority salary scale PO2 (then £12,456 per annum). This compared with the scale agreed for the Outreach/Support/Placement workers of SO1 (then £9,255-£10,228 per annum). The GERU worker was on holiday for the meeting but sent a letter making a number of points including the suggestion that there should be discussion on the possibility of implementing a policy of wage parity, stating

> obviously there would be huge practical and financial implications. Considering, however, the general philosophy behind the new Charlton site, i.e. of general cooperation and shared responsibility between people, I feel it's something the Consortium should at least discuss.

The minutes of the meeting record that the Consortium 'felt that this was something they might like to see happen but it was impractical'.

However no examination of the implications for the proposed 'democratic management structure' was undertaken. The Consortium devised a set of arrangements for the recruitment procedures. Two sub-groups were to be set up to handle the shortlisting and first interviews of applicants for educational/training posts and support/outreach/placement posts. The criteria for shortlisting were to be referred back to 'Group A', which would undertake final interviews and also undertake all aspects of the recruitment of Centre Coordinators. The recruitment procedures were halted because of the problems which arose over the use of Section 45 funding. However when in October the question arose about the composition of the Board of Directors, it was decided by the Consortium that it should consist of the membership of Group A. This original group remained substantially the same, with the substitution of the replacement GERU representative and the addition of the Greenwich Afro-Caribbean Association representative, until the first Annual General Meeting not held until December 1985.

By November 1983 concerns were being raised about the level of involvement of the local community. The need to develop widespread involvement was recognized in the earlier, unsuccessful, application for funding for part-time outreach workers. In a report to the Consortium on 17 November the Development Worker stated that

> It is obvious from Consortium attendance that there is a disillusionment that the process has taken this long and the Consortium will need to put considerable energies into both attracting those who have been involved but who now give the Consortium a low priority and involving new members.

At the previous meeting he had pointed out to the Consortium that his work had changed and that he now did a considerable amount of administrative work. In January 1984 the Consortium decided to apply for funding for two other support workers and for renewal of funding for the development worker. One of the additional posts was intended to be an administrative/finance worker, the other to act as administrative worker/typist. This was approved by GLTB. Also in January the Consortium agreed the need to examine in detail the composition of the Consortium and how to ensure that the membership reflects the positive action policy of Consortium.

The Consortium's Operation of the Centre: Staff Appointed

The first 'permanent' workers were appointed in April 1983, amongst whom was the Trainee Support Coordinator. However the other two Centre Coordinators were not appointed until several months later. Soon after the appointment of the first group of workers, concerns were raised by the GERU representative about the employment practices of the Consortium.

She expressed these in a discussion paper to the Consortium, in particular that

- although the Consortium was by now employing five people the members of the Consortium had not yet begun to think of themselves as employers;
- no one had yet called on the Trainee Support Coordinator, who started work on 2 April, to see if she had started, where she was based, what her working conditions were, how she saw the job;
- similarly the two other support workers recruited had also been 'left to sink or swim';
- having worked in the same office as the Development Worker, she had been horrified at the expectations and demands placed on him.

She concluded that

> the only hope of success we have is to appoint workers who do as we think, and not as we do. ... we are operating line management under the guise of community participation. ... If we haven't been able to support the one worker we have had for over a year, then I'm really fearful of what we can offer to new workers.

The paper was discussed by the Consortium on 11 April 1984 and members 'on the whole agreed' with the points made. Various offers were made for help with dealing with administration, such as typing and photocopying. It was also agreed that the development workers workload would be looked at.

In May 1984 the GERU worker, after discussion with the Development Worker, the Trainee Support Coordinator, and the representative from a local adult education institute, prepared a discussion paper on the 'Role and development of Consortium'. In that paper she expressed the need to consider 'as a matter of urgency' two related issues: the management structure to operate at the Centre and the interim structure and function of the Consortium. Three possible models for the management structure were presented. The first model was that of worker-led line management, the three Centre Coordinators taking the major responsibility for organizing the Centre, evolving their own management structure. A second model was that of the Consortium being a traditional management committee, the workers preparing reports and making proposals.

However the paper urged that a third alternative be adopted, that would 'use the structure outlined in the Constitution and Consortium papers' so that the Consortium would become the vehicle for

- community input into continuing development of CTC;
- policy making body rather than endorsing body;
- 'part of a radical new Management Structure attempting to operate collective ways of working, and worker participation in management of a large institution, and in implementing some of the ideas of popular planning'.

The paper proposed establishing a number of groups to be responsible for intensive thinking about key aspects of the Centre's work, 'think tanks' developing ideas to be presented to the Consortium.

Sub-Committees

However the Consortium did not discuss the paper as the subsequent meeting (on 17 May) finished, without completing its agenda, at midnight after starting at 7pm. Moreover at this stage the Consortium had still not been formally constituted although the Company had existed as a 'corporate shell for several months'. The GLTB representative reported that this was causing difficulties. The Consortium eventually discussed the proposals at a meeting on 31 May 1984, at which the proposal to set up Support Groups to act as 'think tanks' was agreed. Other constitutional issues were discussed and agreement was reached to decide on the composition of an interim Board of Directors and on arrangements for transition to a permanent Board, at the next meeting. Also at this meeting it was agreed that Centre Coordinators would be ex-officio members of the Board, that the staff would elect two temporary representatives, and that by the end of July any member of staff could attend meetings as observers.

At the following meeting, after discussions had taken place between the Consortium's solicitor and certain members of the Consortium, it was agreed to

> delegate powers to the Sub Committees to implement on a day to day basis policies as decided by the Board. Such Committees to include a Finance Committee and other Committees to be decided by the Board. … there should be a minimum of 4 members and a maximum of 10 on the sub committees and … the Convenor of each sub committee should be a member of the Board. … The Board of Directors will appoint/remove the sub committees but the sub

> committees will have the power to invite people in an advisory capacity. The sub committees will report back to full Consortium meetings through Convenors.

At the same meeting the issue of how the Board of Directors should be constituted was debated. The Trainee Support Coordinator proposed some alteration to the existing arrangements as set out in the original Articles of Association. The minutes record disagreement but do not record what decision was reached, if any. The membership, ex officio, of the Board by the three Centre Coordinators was reaffirmed. The minutes also state that there was 'a short discussion on whether the Coordinators were responsible for managing the Centre, and whether the staff were responsible to the relevant Coordinator'. The minute concludes tersely: 'This was not resolved'.

Three weeks later the Trainee Support Coordinator presented to the Consortium a 'Discussion Paper on the Charlton Training Consortium's management structure and its relationship to the Centre Staff'. At this time the Trainee Support Coordinator was the only one of the three Centre Coordinators actually to have started. The Training Coordinator had been appointed but had not yet started, as he was awaiting notification of acceptance for voluntary redundancy by his then current employer. An Administrative/Finance Coordinator had yet to be appointed.

The Trainee Support Coordinator pointed out that the Consortium had agreed with the view expressed by the GLC that the existing structure was inadequate for day-to-day management of the Centre. The task was therefore 'to construct a structure that can ensure effective accountability of the workers to the Board (community) on the one hand and a highly efficient smooth running training centre'. She asked whether staff would be responsible to the Board via the working groups (also referred to in the Consortium's documents as 'Support Groups' and 'Sub-Committees'), in which case the mechanisms for this to happen needed to be stated, or whether it was to be assumed that the Centre Coordinators would be responsible to the Board for staffing matters.

By the latter she meant grievance procedures, internal problems between trainers and trainees, and any problems requiring on-the-spot decision. She indicated that the higher salary and generalized work indicated an implied responsibility on the part of Centre Coordinators. She argued that

> it will greatly affect the efficiency and morale of the Centre if decisions of a relatively minor nature cannot be conducted by the Centre Coordinators, as they are the people who will be responsible for the smooth running of the Centre.

She concluded:

> By way of summary I would like the Consortium to take on board and act on the following points:
>
> 1. a) clarify lines of responsibility between Centre staff and the Board of Directors,
>   b) by what method this is to be done;

2. a) clarify how staff will be accountable to the Board,
   b) by what method and through what mechanism this is to be done;
3. Decide upon what decisions will be delegated and to who;
4. As a result of the decisions made on the above, re-assess the Centre Coordinators' responsibilities and make them more explicit.

At the 28 June 1984 meeting of the Consortium, the Trainee Support Coordinator's paper seemed to result in little discussion. The minutes record the item in just three sentences, finishing with: 'it was decided that workers should do papers on how they saw their roles and bring back to a meeting in two week's time'.

In a letter to the Chair of the Consortium, an officer with Greenwich Council, the representative of the local Racial Equality Council who had actually chaired the 28 June meeting in the Chair's absence, expressed her feeling that 'there were no satisfactory conclusions to any of the Agenda items'. She stated some of her 'areas of concern' including her views on the issue of 'lines of authority' saying 'I remain unconvinced that working groups could be expected to take responsibility for the day to day running of a Centre the size of Charlton'. She stated that she believed that the Board should appoint a 'Management Group' to manage the Centre, with the three 'top posts', the three Centre Coordinators, being responsible as a collective to that Management Group. The Chair of the Board should be Chair of the Management Group, and the existing working groups should be seen only as support groups to workers. She concluded: 'I would expect the three workers to take full charge of their areas of responsibility and be responsible as a collective for the decisions at the Training Centre on a day to day basis'.

The meeting two weeks later, 12 July 1984, agreed to defer discussion to the first item at the next meeting, although one member asked to be minuted as objecting because the item had been deferred so many times before. The next meeting did not discuss the paper, which was eventually discussed at the meeting on 27 July 1984. At the 27 July meeting the GERU representative tabled a diagram indicating her views on the lines of accountability/responsibility. The minutes for the meeting contain a note from the minute taker that 'this was a very long item, with most members present contributing to the debate, below is a list of the main decisions taken, rather than a resume of the discussion'. The minutes record the decisions of the meeting as:

> Role of Centre Coordinating Team: Agreed that on day to day basis this team had the responsibility of ensuring smooth and efficient running of the Centre. …
>
> Women's Unit: Agreed to delete phrase 'semi-autonomous' from description of women's unit. Unit is an integral part of the Centre, although different from other units because it has more than one function – i.e. has training, and trainees support. The women's unit relates to all 3 coordinators, the women's unit coordinator has the responsibility to ensure that Centre coordinators are fully informed of all the activities within the Unit.

> Sub-groups/working groups: Agreed to adopt sub-group structure as previously discussed. Development sub-group to look at terms of reference.

The GERU representative recalled that the paper she presented was seen as an attempt to get the Women's Unit Coordinator on the same level as the three Centre Coordinators. This was seen in part as racist, since it was seen to be 'putting down' the Trainee Support Coordinator, who was black. Much of the discussion was concerned with the fact that the diagram presented had the Women's Unit in the centre of the page, and this was interpreted as intended to place the Unit in a central position. The Trainee Support Coordinator recalled that the discussion of her paper 'didn't get a proper hearing; the discussion got sidetracked onto the question of the Women's Unit. The Board was possibly going through a process of discussing differences of ideology, issues of accountability of the Women's Unit'.

At the following meeting of the Consortium, 3 August 1984, proposals regarding the composition and responsibilities of the Sub-Committees were agreed. These were detailed in a paper which pointed out the various issues which the Consortium had been discussing for some time and went on

> Given that it is now of urgent importance that the matter be clarified, the following is set out as a way of combining the respective desires to ensure that the Board determines policy, that coordinators have a distinct role to play, and that all members of staff have jobs that involve significant degrees of responsibility in line with the Consortium's desire to upgrade the quality of jobs wherever possible.

It was agreed to set up nine sub-committees:

- Finance/Administration;
- Staff;
- Trainees;
- Training;
- Education;
- Childcare;
- Women's Unit;
- Consortium Development;
- Development of the Centre.

Outline terms of reference were proposed for each sub-committee, which each had responsibility for developing policy proposals, for developing systems to implement policy, for overseeing implementation of systems, and for monitoring. The meetings of sub-committees would be the basis for overseeing the work of the three Centre Coordinators.

> On the basis of these meetings, the Coordinators would be responsible for day to day management of the Centre in line with the Board's stated aims and policies. ... Clarification is needed of which staff are responsible to which Coordinator. ... The Coordinator would be responsible for day to day management and would be responsible for delegating work in a manner consistent with cooperative working and ensuring maximum responsibility and job satisfaction.

The minutes of the meeting record that 'it was agreed that there was a great deal of business to be carried out over the next few weeks and that this would necessarily involve the sub-committees meeting a great deal in this period'. The meeting which agreed this was attended by 8 people, including the Trainee Support Coordinator, the Training Coordinator, who had just started, and one member of staff.

Relations Between Board and Staff

During the period between April 1984 when the first workers were taken on, and October relations between the Consortium Board of Directors and the workers became strained. At a meeting on 1 October, called for by the workers, they described to the Board the complaints they had, including

- having to justify everything they asked for, including the most obvious and trivial;
- feeling that their knowledge and experience were not respected;
- and that they were given no autonomy;
- having to battle for acceptance of their ideas just as they would in a hierarchical workplace;
- having no departmental budgets;
- making recommendations in sub-groups which didn't even get discussed by the board;
- not knowing how much power is vested at each level of structure;
- not having an administrative/finance coordinator, who should have been here from the beginning.

Until this time staff representatives at Board meetings had not been voting members; it was clarified at this meeting that representatives were full members of the Board. Apart from this clear decision, there appears to have been little practical outcome of the meeting. The work of the Centre continued, and some courses were started by December 1984. Difficulties over building plans caused problems over the intended childcare provision. Issues about staff contracts continued to be discussed, as were the details of disciplinary and grievance procedures. Staff continued to experience difficulties in obtaining resources such as typewriters, books, and so on. Clerical support was, at this stage, being provided by workers employed by Outset, an organization specifically set up to employ workers with disabilities. However, problems arose, largely because the necessary adaptations to the buildings could not be carried out. This caused difficulties for staff wishing to have reports, handouts, and training manuals typed.

Reorganization

The three Centre Coordinators were asked to prepare proposals for the organization and management of the Centre, for discussion at a special Board meeting on 26 April 1985. In fact it was the Trainee Support Coordinator who prepared the main paper with the other two Centre Coordinators presenting short papers commenting on those proposals.

The paper presented by the Trainee Support Coordinator referred back to the paper presented in June 1984. She stated that discussion which took place around that paper went only part way in clarifying lines of responsibility and left other questions undecided, stating that 'undebated questions represent many blind spots which are having a negative affect on the effective functioning of the Centre at present and will continue to do so in the future'. She pointed out the difficulty in developing a structure for an organization the size of Charlton, in line with what the Board had stated it wanted to see.

> It seems to me that there was no clear guidelines laid down by the Consortium around the structure of the Centre except that it should have a cooperative working structure. I have not, so far, observed any examples to draw comparison or detailed instructions around how this should evolve or work. ... The problem at the moment is that Charlton Training Centre is unique with apparently no similar model to look at. ... The quest for a new imaginative, alternative structure for the Centre is in fact slowing down the pace of development.

There was considerable discussion about 'ideology', 'perspectives', '-isms'; 'racism', 'sexism', 'heterosexism', 'class', 'collectivism', 'cooperativism' were cited as examples, and differences of perspective among these were mentioned. The repeated discussions about the Women's Unit were cited, as was discussion about discipline, i.e. when issues were raised about timekeeping of trainees some workers would state that the 'did not believe in social control'. The Administrative/Finance Coordinator stated that there were difficulties created by the debate between 'collectivism' and 'accountability' in terms of financial control procedures and personnel procedures. One of the Board members argued that the debates over ideological perspectives should not affect the training which should be regarded as a priority: trainees were there to get skills and would have to go out to get jobs on that basis. The Chair endorsed this view, which was not opposed, or, at least not openly. The Trainee Support Coordinator presented a proposed framework within which every Centre worker was responsible to a designated

Centre Coordinator. The Board agreed with this and decided that 'no action or request would be considered unless it came through the Coordinator'.

During the presentation and debate the proposals being made led to several members of the Board stating that these were exactly what they understood to be the agreed situation and expressing surprise that there was any confusion or disagreement. The Trainee Support Coordinator stated that problems had arisen because the authority of Centre Coordinators and the lines of responsibility were unclear; in practice if a member of staff disagreed with a Centre Coordinator there was little the Coordinator could do. Several members of the Board, in particular the Chair and the Board Secretary (GERU representative) reported on the frequency with which members of staff would contact them to sort out a problem, a practice which would cease under the new arrangements.

In a discussion shortly after this meeting, the Trainee Support Coordinator expressed concerns about the way in which the Board had dealt with the issues. She stated that she did not feel that Board members clearly understood the issues about the formal structure and the actual practices of people. She added 'They kept saying "That's what we thought was already supposed to be happening" but the issues have been raised many times before without any results'.

The next Board meeting agreed to put onto the next meeting's agenda the discussion of the special meeting regarding the structure. The next meeting was held five weeks later at which it was resolved 'that this paper be brought back to the next Board meeting to enable Centre workers to examine the proposals'. The Trainee Support Coordinator stated that she 'wished to have it recorded that she hoped that these papers would be considered and a decision arrived at because of the great effort involved in preparing it for discussion'.

She resigned a few days later; the paper on the structure was not discussed after that time.

Changes in Administration

During June 1985 the Chair appointed a worker employed by Outset (the organization providing clerical support) at the Centre as Personnel Officer. She was away at the time of the next meeting of the Board when worker representatives objected that the equal opportunities procedures had been breached. Because no job description existed for the post, and it had not appeared in any funding application, the meeting decided that the appointment was invalid. At the subsequent meeting the Chair justified her action on the grounds that there were severe problems with regard to staff matters and that as Chair she was responsible to ensure that the Centre ran smoothly. Because she was about to go on a trip to America she was compelled to take 'Chair's action'. The voting on the matter was split and the issue was not raised at any subsequent meeting.

Also at this meeting a member of the Board expressed concern about the tendency for workers to regard the Board as just the community representatives.

However when the worker representatives, Centre Coordinators, and other workers had attempted to enter the meeting room they were asked to wait outside while the Chair, Vice Chair, Treasurer, Secretary, and GLTB representative finished a meeting they had been holding. The workers' representatives raised this matter, to be told that the prior meeting was a meeting of honorary officers. A workers' representative responded to this that 'You've made it crystal clear that there is a division between the officers of the Board and the representatives'.

Problems about finance and administrative matters were raised at the next meeting. The Administrative/Finance Coordinator had for some time claimed that his workload was excessive, that the arrangements with Outset were not working well, and that extra staff were needed in his department reporting to him. He was asked to draw up detailed proposals for the next meeting. However at the next meeting he was dismissed. No worker representatives were present, and the one trainee representative was asked to leave the meeting. A member of the Education Unit staff was appointed Acting Administrative Coordinator, and the Treasurer was engaged to work at the Centre, for an honorarium, to deal with the financial administration.

Analysis

The above recounting of the story of the Charlton Training Consortium displays the considerable problems encountered in attempting to develop a democratic approach to organization and management. A number of issues arise.

Lack of Clarification

The idea that Charlton should be organized and managed in some form of 'democratic', 'non-hierarchical', 'collective' manner existed from the very early stages. However there was little work done on clarifying the meanings of such terms, or on establishing how such ideals might be put into practice. Certainly during the 'development' period (from July 1982 when the first meeting of 'interested parties' took place until December 1983 when the funding application was approved by GLTB) there appears to have been no overt questioning of the principle of some type of democratic form of management within the Consortium. The GCRE representative did raise concerns in a letter in July 1984 to the then Chair (Greenwich Council representative) but there is no record of any debate on the matter in other available documents. The accepted view that the development of internal management structure should and could be undertaken when workers started is reflected in the concern of the first set of workers not to develop any detailed arrangements until the majority of staff had been employed.

There existed then no agreed arrangements by which decisions could be arrived at about how the structure might be developed. Within this 'structureless' situation we can see developing the kinds of informal structures which Freeman describes as examples of the 'tyranny of structurelessness' (Freeman 1972). The tendency for an 'elite', an 'in' group, to form is described by the GERU worker who had been one of the main initiators, who recalls the difficulties of gaining involvement of local people:

> Some members made efforts to bring others along with them. But the meetings were regularly inaccessible because workers used jargon, because the agendas were too long. ... Members who used manipulative meeting tactics were those largely with experience in or knowledge of councils or politics. ...We didn't give enough status to isolated individuals who for different reasons couldn't or wouldn't get involved in groups or meetings, i.e. often the exact people we wanted to gear the skill centre towards. (Mantle 1985: 8)

Various members of the Board were called upon on many occasions to intervene in the Centre on particular matters, without reference to any official lines of decision making. By responding to such calls the informal structures were reinforced. Similarly, because no official mechanism for representation of workers on the Board was developed, individual workers were able to attend Board meetings to attempt to influence decisions. This then undermined the nascent form of representation being developed, and representatives resigned as such and others refused to accept the post of representatives.

Democracy versus Efficiency/Effectiveness

The events which occurred illustrate a major issue for organizations attempting to develop democratic structures of control, that of how to meet the aims of efficiency and effectiveness and at the same time to promote the widest possible participation in decision making and control. These are usually seen as needing to kept in some kind of 'balance', as being potentially conflicting. One main area of potential conflict was concerned with the time involved in consultation, discussion and decision making. This was referred to in respect of the length of meetings which rarely end with the agendas completed. Another area of potential conflict was concerned with the issue of unequal distribution of expertise. It was clear from the way in which the staffing arrangements were planned that the Centre Coordinators were expected to have considerable managerial expertise individually and collectively, and the proposals for reorganizing put forward in April 1985 would, in theory, have provided for Centre Coordinators to have more authority in decision making.

A third and perhaps most critical area of potential conflict lay in the differing perceptions participants appeared to have on the very nature and purpose of the Centre. Although some recognition was given to this, such as references to '-isms' and 'ideologies', this was in practice generally not tackled explicitly. The (second) GERU representative described the tendency to avoid such discussion about differing perceptions as a naive belief in 'unity through oppression', that there would be a common bond because of the experience of oppression. It is difficult to ascertain why the potential conflict of perceptions over the nature and purpose of the Centre was not dealt with explicitly. However it is not difficult to understand why this might happen. The issues which the Consortium and the Centre were tackling are of great personal and political importance.

The nature of many participants' commitment was a moral and political one. But the bringing together of these participants was no guarantee that there will be unanimity about the directions which should be taken. In fact there were numerous ways in which conflicts had arisen. The Women's Unit faced problems over the specific way in which the involvement or non-involvement of men should be tackled, in which the charge of racism was laid against those who, in seeking to exclude men, were seen to be dividing black women and men.

Disagreements arose over the nature of non-vocational activities. Educational support linked to training was recognized as important, although there was a difference of view about whether or not this should be 'voluntary' or 'compulsory', and if the former how to deal with problems of 'stigma' and absence from the workshop. There was an even greater difference on the nature of activities which MSC would refer to as 'life and social skills training' (Moore 1983). For some workers the time set aside for this, one afternoon per week, was an opportunity for 'consciousness raising' and political education. Others, including trainees, considered that the more pressing problems of obtaining full welfare benefits and job finding should be dealt with. Although there was debate and discussion on these and other issues, in the absence of any agreed or recognized and enforceable decision making structure the issues were never fully resolved.

Underlying Structure

However, it would be a mistake to conclude that there was an absence of structure at Charlton. The power within the Consortium was clearly weighted heavily among those who are not employed in, or trainees in the Centre. These had all been involved with the Consortium since at least early 1983, and represented organizations which have been involved since the beginning. For several meetings, until the matter was raised by a worker, representatives were not in fact voting members. The Centre Coordinators did not vote at meetings, although it was agreed that they should be full members of the Board. Technically, the Centre Coordinators were not directors since they were not registered as such, but then neither were worker and trainee representatives.

The Articles of Association provide for a reorganization of the Board at the first Annual General Meeting, at which time the Centre Coordinators and representatives legally become members of the Board of Directors. The AGM was postponed at least twice and the Consortium had been incorporated for longer than 18 months which is supposed to be the time within which the first AGM was to be held. Moreover it was clearly the intention of various Consortium and Board meetings that the Coordinators and representatives *should* be full members of the

Board, even to the extent of being able to vote on matters relating (in the case of employees) to their own contracts of employment. There was then a division between those who were employed or trainees at the Centre on the one side and other (community group) members on the other.

Moreover, as we saw in the previous chapter, much of the crucial decision making was being made by the state agencies involved. The Board, and other groups and individuals at Charlton were, in effect, working within the constraints established by the policies and strategies of those agencies. Decisions about staffing required and the salary levels were made in the context of applications for funding. Such salaries were based on local authority scales; the terms and conditions again were based on local authority guidelines. When an alternative, pay parity, was suggested this was rejected as 'impractical', and so a hierarchy of salaries was introduced at an early stage. Later, the very fact that Centre Coordinators were paid more was used to justify the proposition that they should have considerably more responsibility than others. At the meeting in April 1985 to discuss the structure some Board members commented that the structure proposed by the Trainee Support Coordinator was 'what is supposed to be happening already'.

A distinction was made between the 'structure' and 'practice', and it was agreed that a training programme be set up for staff to enable them to work in the structure agreed. It was quite specifically agreed that such training was not to be concerned with changing that structure. In this way 'structure' was reified, rather than treated as the manifestation of a particular patterns of social processes. Moreover, such structure was only related to the staff and trainees in the Centre; the Board and Consortium were considered to be separated from the Centre's management structure. Yet the members of the Board were actively involved in the day to day running of the Centre. The Treasurer's approval was required for all expenditure, and routine items such as stationery and essential items such as course books became difficult to obtain.

Although the staff raised such matters with the Board in October 1984 only minor relaxation of this practice occurred. It took several months of requesting, arguing, etc. to have typewriters ordered. The Chair was often called to deal with disputes, especially over matters of expenditure. The GERU representative, who was Company Secretary, was also often called to intervene usually over issues of staff relations and the Women's Unit. By responding to such calls the 'off-site' Board members reinforced their dominant position. The Board was pointedly divided between such members, and the worker and trainee representatives.

Control, Participation and Ideology

The continuing lack of clarity about the way the Centre was to be managed, the repeated and ineffective reconsideration of the structure, and the actions of the Board members all served to prevent the development of a democratic organizational form. The emphasis in the discussions on the structure was on authority and control, concerned with issues of issues of efficiency and economy.

In this they reflected the dominant issues for management in capitalist enterprise. Although the staff objected on occasions, there was no successful strategy for changing the nature of relationships between the Board and the staff. This was due to the fact that the major decisions were in reality being made outside the Centre, and the Board and staff engaged in marginal aspects of decision making. Thus they became focused around matters which were regarded as technical and administrative in character.

The political implications of these affairs, that is, that the Centre was not actually operating in a democratic manner, were rarely raised. Indeed they were discounted as examples of '-isms', and the purpose of the Centre was continually reiterated as being to train the trainees. The opportunity provided by the Centre's existence to develop a democratic organization, with consequent implications for the nature of management in such an organization, was thereby lost. Insofar as those involved perceived the important issues to be about meeting the 'productive' aims of the Centre (training unemployed people), or in effect allowed this perception to take dominance, the conventional ideology of organization and management took material form. However, this was always subject to the espoused ideology of democracy.

So the Centre continued in a constant state of uncertainty and tension as the underlying contradictions worked themselves out in the day to day events. No effective challenge was made to these contradictions, and so no transformation of the organization took place. The main lesson from this unfortunate episode is that the conventional ideology of organization and management, concerned with control and authority, and the ideology of participation and democracy, must be examined by any group or organization espousing the latter, in order to ensure that such democratic ideals are not subordinated to the perceived requirements for efficient control.

Chapter 12
Beyond the Dominance of Management?

The previous four chapters have presented 'thick descriptions' (Geertz 1973) of the two case study organizations. Each had started with overtly espoused radical intent, based on understandings of participatory approaches to their organization and management. Each was transformed, over the period studied, to more conventional modes of hierarchical, non-participatory management, excluding the voices of the constituencies they had originally intended to be included and engaged in the planning, decision making and execution of the projects for which they were established. In this, they were and are not unique (see, for example, Messinger 1955, Sills 1957, Zald and Denton 1963, Banfield and Hague 1990). The time period of the events depicted is now a quarter of a century ago. Both organizations were based in a 'developed world' context, in London, a major metropolis and centre for international finance and trade, as well as host to the Olympic Games 2012. The concerns of these two organizations might seem marginal to the concerns of economic and social development in the early twenty-first century. In this chapter, we shall consider some issues arising from the foregoing study, to identify the continuing relevance of the study.

Telling the Stories

A first key point to make here is that of the importance of 'telling the stories'. We might consider here two sayings, albeit somewhat hackneyed: Santayana's aphorism that those who cannot remember the past are condemned to repeat it, and the saying that history is written by the winners. In the two cases examined here, we should note that the circumstances in which they arose continue to affect communities in inner urban areas not only of London but of other cities in the developed world. The costs to governments of failure to attempt to address such problems are very high, particularly in the aftermath of the financial crisis and economic recession of the latter part of the first decade of this century. The UK, like many other western governments, faces major fiscal difficulties arising from the 'bail-out' of financial institutions deemed too important to allow to fail, lest such failure trigger a 'meltdown' of the global financial system. Governments, national and local, will therefore be squeezed between such fiscal limitations and the likely costs, economic, social, political, of failure to take action of the problems facing communities such as those involved in the Bus Garage and Charlton projects.

It is almost certain that 'community-based' and 'participatory' initiatives will be viewed by government as a key instrument for delivering solutions at relatively low cost. The funding framework for such initiatives will be that set by national government, using its legislative and fiscal powers, as we have seen in the case of the two projects studied. The specific local dimension of any initiatives will, however, be subject to the support or otherwise of government locally, always subject to the national government legal and funding framework but as mediated by the political orientations of the party in control of the local government machinery. The local requirements may be consistent with those set nationally or, as we have seen, may be inconsistent or oppositional. Poorly resourced groups seeking funding for the initiatives they develop, to meet the needs of their communities, will be constrained by the requirements set by funders, national and local.

The dynamics of such funding mechanisms are therefore likely to lead to groups moderating the radical nature of their original definition of the problems they seek to address, and of their intended strategies for dealing with them, in order to obtain funding. They will be constrained to define the problems and propose strategies for action that align with those set by the funding agencies. Moreover, this will occur not just at the point of initial funding application but throughout the life of the initiative, through reporting mechanisms and funding continuation procedures (Holmes and Grieco 1991). The events that unfolded in our two projects seem destined to be repeated; unless that, is, the stories of the two projects may form the basis on which the instigators of new initiatives, and those that support them, may be in a position to 'remember the past' in order to avoid condemnation to repeat.

Moreover, one reason why it may often be the case that history is written by the winners is that the 'losers' do not tell their history, or at least, do not do so in a form that affords a widespread audience. That is, 'winners' and 'losers' are creations of the dynamics of (hi)story-telling. By telling the alternative story, or stories, to that presented by the dominant voice of funding agencies, the winners-losers relationship may significantly change and even dissolve. However, until very recently, the ability of poorly resourced groups to present their own stories in a manner that afforded access to a widespread audience has been severely limited. The development of new forms of information and communication technologies has radically transformed that situation. We shall examine this development later in this chapter.

Beyond Binarism: Globalizing Alternatives

In her study, within this book series, of the employment relationship based on ethnography of two lock manufacturing companies in the midlands of the UK, Greene argues for the need to break away from 'rigid binarisms between the developing and the developed world, traditional and modern, new and old' (Greene 2001: 119). Typically, she argues, even where studies are undertaken of industrial relations in developing countries, the analysis is premised on the notion of what those countries can learn from the experience in Britain, North America and Australia. She argues, in contrast, for more attention to what lessons may be learnt by the developed countries from the examples of developing countries.

We might extend such a call to break from such binarism by remembering that even in the developed world there are continuing areas of economic and social deprivation, whilst in developing countries there are areas of affluence. The globalized nature of capitalism, and of the division of labour, makes the use of such binarism simplistic and misleading. Communities and groups in both developing and developed world contexts will seek ways to overcome the disadvantage, deprivation, poverty, domination and exploitation they experience. Of course, particular circumstances, in particular situations, may vary enormously. A common feature of such initiatives will be that those instigating them will be relatively poorly-resourced. Where there is the opportunity to apply for relatively large-scale funding support from a major government or inter-government agency, such groups may see such opportunity as one that will bring great benefits without considering likely consequences.

One possible response is to resist such temptation and attempt to develop an initiative from the group's own resources, supplemented occasionally by resources granted freely and without strings. Except in very limited situations, almost divorced from the economic, social and political context, the limitation of resources make this option very difficult (Grieco 1998). However, there is more than one other option than that of accepting support from an agency whose modus operandi is contrary to that of the group. There are a number of supporting agencies and associations that are themselves based on and supportive of participatory principles.

A famous agency that provides funding support is that of Grameen Bank of Bangladesh (Yunus and Jolis 1998, Dowla and Barua 2006). This is just one example of a range of approaches to microfinance, providing funding to low-income groups to enable them to seek ways out of poverty (Ardener and Burman 1996, Grieco 1998, Organisation for Economic Co-operation and Development 2003, Matthäus-Maier 209). The participation and involvement of the users of such services is, according to Grieco (1998) a key element in the social enforcement of repayment of loans taken, and the structures for such participation will vary between locations and groups.

Resources, in themselves, will not be sufficient and, as we have seen, may become a diversion or even a trap, unless the analysis of the 'problem' is appropriate. Typically, community-based initiatives start from immediate experience of the situation, and those who instigate them will have little or no knowledge or awareness of other initiatives by other groups and communities. Supportive structures may therefore be of considerable help if they can bring access to relevant expertise without imposing particular modes of understanding or prescriptions for action that may not be suitable and/or may interfere with and undermine participatory forms of organizing.

One example of such supportive structure may be that of the trade union movement, as can be seen in respect of Self Employed Women's Association

(SEWA) in India, a trade union established by and supporting poor, self-employed women workers (Rose 1992, Datta 2003). On a larger scale, we may note the example of La Via Campesina, an international movement based in Honduras that coordinates peasant organizations (small and medium scale producers, rural women, landless movements, indigenous communities), in Asia, Africa, Europe, and the Americas. It aims to develop solidarity and unity among such organizations, to promote gender parity, fair economic relations, food sovereignty and sustainable agriculture, and to preservation natural resources such as land, water and seeds (Mander and Tauli-Corpuz 2006: 241).

An ever-present danger is, of course, that organizations engaged with such groups and communities, and those groups and communities, may *not* have a common understanding of the issues to be addressed or of the manner in which this may be done that promotes participatory organizing and action. Or, even if they do have such shared understanding, the nature of the 'expert–client' engagement may undermine the capacity for such participatory form (Nelson and Wright 1997, Dar and Cooke 2008). There is a need for amplification of the voices of the instigators and participants of such community-based initiatives.

Amplifying Voices: The Potential of ICTs

It is now well-recognized and accepted that the development of information and communication technologies over the past decade and a half has seen a radical transformation modes of communication. The rapid growth of the world wide web from the mid-1990s, coupled with mobile phone technology, have undoubtedly wrought major changes in modes of communication that far outstrips those of the invention of moveable type printing. Quite how these changes may best be understood, and what they portend for society, the economy, politics, and everyday life have been the subjects of debate (Wilhelm 2000, Castells 2001, Webster 2002, Wellman and Haythornthwaite 2002). Castells argues the potential social outcomes of the new technology cannot be predicted: it is 'to be discovered by experience, not proclaimed beforehand' (Castells 2001: 5). Here we shall consider the opportunities afforded by for disparate and poorly-resourced groups to communicate with each other and to the wider world.

A key affordance (Norman 1988) of the new communications technologies is that of reducing the effects of spatial separation on the coordination of activities between individuals and between groups: what has been termed the 'death of distance' (Cairncross 2001) or 'electronic adjacency' (Little, Holmes and Grieco 2001). This came to widespread public awareness by media coverage of the way the technology was used to great effect by anticapitalist, antiglobalization activists organizing a 'Carnival Against Capitalism' in the City of London on 18 June 1999 (Pickerill 2006). Whilst commercial business has increasingly made use of the web for its own purposes, the network infrastructure has afforded activist groups of various kinds to disseminate its own views. The set up and transaction costs are relatively low when compared with those required for mass production and dissemination of information using previous technologies such as print. Whereas governments have, in the past, been able fairly easily to disrupt dissenting groups, by locating central coordination points and destroying (relatively expensive) equipment, the internet and technologies required for its access have transformed the balance of forces (see, for example, Holmes and Grieco 2001).

Another key affordance of the new technologies is that of multimode presentation: text, image, audio, and video can all be readily combined into single or multiple 'pages'. This helps to overcome the bias towards textual documentation that may be seen in the two initiatives studied in this book, particularly the Bus Garage Project. Not only did such documentation place demands on members of the Community Council who, although clearly literate and highly intelligent, were unpractised in dealing with the dense and formal prose of reporting produced by the funding agencies and also required of the Steering Group by those agencies. Furthermore, limited funding prevented widespread communication to the local community, this being mainly confined to very occasional newsletters and public meetings. A quarter of a century later, the advances in ICTs would enable a similar group to have a steady stream of communication with the local community in modes that are attractive and readily accessible.

One major area where this would have helped is in respect of the plans for the site, whose physical features presented an obstacle for community engagement. As it was a building site, safety and security considerations prevented open access. The local community could see little to gain a view of progress being made. Moreover, the plans for the redevelopment were physically manifested in then-traditional architectural forms of drawings and a balsawood model. Modern, computer-aided approaches to architectural design would enable such plans to be made available online. The graphical features of such programs would enable realistic 3-D visions of the proposed new centre to be displayed, even permitting virtual journeys through the virtual centre (Holmes 2003). Access to this facility could be provided through public access internet-connected computers, in libraries, schools and community centres, available at any time and rapidly updateable.

Such *dissemination* affordances would play a part in maintaining engagement by and support from the constituency for whom a community-based initiative has been established. A further affordance is that of *interactivity*, that is, enabling active engagement between ordinary members of the community and the activists who instigated and are engaged in the community-based initiative. This would enable polling of preferences on options facing the project; support and help may be solicited; members of the community would be able to pose questions, to express opinion, to offer help. Rather than rumour, ill-informed criticism and discontent arising from limited communication, the new technologies could enable informed discussion and debate within the community, helping to develop and maintain positive relationship between 'leaders' and the community they seek to serve.

The new technologies also enable *recording and archiving*, preserving the memory of intentions and of events. As we have seen, in both the Bus Garage and the Charlton Training Centre projects, the founding values were subject to slippage as various pressures and demands were placed on them. Although they were matters of record, the recording was in documents not readily available as reminders to those who were involved, as founders or as officers of espousedly supportive agencies. More importantly, they were not readily available to new entrants onto the scene or to members of the local communities. There were critical points at which major decisions affecting the future direction of the projects were at stake; the new technologies might have afforded ready access and so brought more explicit focus on the implications of each of the range of choices for maintaining or departing from the founding understandings and values. Under similar circumstances now and in future, community-based groups may be able to deploy the technology in a way that serves to limit the tendency for goal displacement and goal succession.

Moreover, archiving of histories is enabled by the new technologies, in ways that completely transform previous possibilities (Green, Grieco and Holmes 2002). The web is now routinely used as a basis for providing multimedia accounts of the histories of events and groups. This facility may be of considerable value in dissemination of participant accounts of community-based initiatives, to counteract the accounts presented by funding and other dominant agencies. In this way, the storytelling that Denning (2000) argues is vital for 'igniting action' in development contexts may be enabled, ensuring the amplification of authentic voices of participants in such action.

Finally

In undertaking the foregoing study we have obviously moved some distance from the focus of conventional management theory. This, as we have seen, is based on a set of assumptions about the nature of organizations, ignoring many crucial aspects of the wider economic, social and political order. Contrary to this, we have worked from a framework for analysis which starts from this wider context, locating our study within that totality. The field studies indicate how important this is for an adequate understanding of the failure of such innovative organizations examined to be successful in tackling the major problems they sought to address.

Thus a first conclusion we may draw is that 'management' should not be the primary focus of analysis. A considerable amount of writing on research on management (and writing not based on research, but repeating conventional prescriptions) treats the nature of management as a taken-for-granted concept, and deals with the activities of managers in their own right. On the contrary, we have seen that the activities of those who occupy 'management' positions often have meaning only in relation to in the wider societal context.

A second conclusion is that the notion of management as disinterested, neutral arbiters between differing sectional interests must be rejected. Management cannot be such for it both arises out of fundamental conflicts between the opposing interests of capital and labour, and is caught up in such conflict being situated in a contradictory class location.

A third conclusion is that any agency engaged in a major intervention which is intended to make fundamental changes economic and social circumstances of low-income and deprived communities, must as a matter of priority examine critically the nature of the organizational and management implications. In particular, the social processes which reinforce conventional patterns of organization and management must be countered by counter-processes to promote democratic participation. Where such interventions are specifically related to the operations of the labour market, through job creation and training schemes, the processes which will tend to reinforce and promote capitalist relations of production need to be identified and countered.

Finally, we have noted how the circumstances under which the two organizations studied, the Bus Garage Project and the Charlton Training Centre, differ significantly from those facing community-based development initiatives currently and the future. The developments in information and communication technologies afford modes of interaction and engagement, for coordination, for dissemination of accounts, at relatively low cost for entry and transaction. The dominant managerial voice, claimed and exerted as authoritative, may yet be countered by the amplification of the voices of the poorly-resourced, low-income, deprived groups and communities previously excluded from influence and decision making on key social, economic and political matters affecting their lives.

Bibliography

Abbott, A. 1988. *The System of Professions*. Chicago: University of Chicago Press.

Abercrombie, N. and Urry, J. 1983. *Capital, Labour and the Middle Classes*. London: Allen and Unwin.

Alvesson, M., Bridgman, T. and Willmott, H. 2009a. Introduction, in *The Oxford Handbook of Critical Management Studies*, edited by M. Alvesson, T. Bridgman and H. Willmott. Oxford: Oxford University Press, 1-26.

Alvesson, M., Bridgman, T. and Willmott, H. (eds) 2009b. *The Oxford Handbook of Critical Management Studies*. Oxford: Oxford University Press.

Alvesson, M. and Willmott, H. (eds) 1992. *Critical Management Studies*. London: Sage Publications.

Alvesson, M. and Willmott, H. 1996. *Making Sense of Management: A Critical Introduction*. London: Sage Publications.

Andreski, S. 1972. *Social Sciences as Sorcery*. London: Deutsch.

Appelbaum, R. 1978. Marxist method: structural constraints and praxis. *The American Sociologist*, 13, 73-81.

Ardener, S. and Burman, S. (eds) 1996. *Money-Go-Rounds: The Importance of Rotating Savings and Credit Associations for Women*. Oxford: Berg.

Aronowitz, S. 1978. Marx, Braverman, and the logic of capital. *Insurgent Sociologist*, VIII(Fall), 126-146.

Atkinson, J. 1985. *Flexibility, Uncertainty and Manpower Management*. Brighton: Institute of Manpower Studies.

Atkinson, M. 1971. *Orthodox Consensus and Radical Alternative: A Study in Sociological Theory*. London: Heinemann.

Banfield, P. and Hague, D. 1990. The Manor Employment Project: The conception, development and transition of a community-based employment initiative, in *New Forms of Ownership*, edited by G. Jenkins and M. Poole. London: Routledge, 63-80.

Baran, P. and Sweezy, P. 1968. *Monopoly Capital*. Harmondsworth: Penguin.

Bates, I., Clarke, J., Cohen, P., Finn, D., Moore, R. and Willis, P. 1984. *Schooling for the Dole? The New Vocationalism*. Basingstoke: Macmillan.

Benington, J. 1986. Local economic strategies: paradigms for a planned economy? *Local Economy*, 1, 7-24.

Benn, C. and Fairley, J. (eds) 1986. *Challenging the MSC on Jobs, Education and Training*. London: Pluto.

Benson, J.K. 1977. Organisations: A dialectical view. *Administrative Science Quarterly*, 22(1), 1-21.

Benson, J.K. 1983. A dialectical method for the study of organisations, in *Beyond Method: Strategies for Social Research*, edited by G. Morgan. London: Sage Publications.

Benyon, J. (ed.) 1984. *Scarman and After.* London: Pergamon.

Berle, A. and Means, G. 1932. *The Modern Corporation and Private Property.* New York: Macmillan.

Beynon, H. 1973. *Working for Ford.* London: Penguin.

Boddy, D. 2002. *Management: An Introduction.* 2nd Edition. Harlow: Pearson Education.

Boje, D. 2001. *Narrative Methods for Organizational and Communication Research.* London: Sage Publications.

Bourn, J. 1979. *Management in Central and Local Government.* London: Pitman.

Bradley, H. 1986. Work, home and the restructuring of jobs, in *The Changing Experience of Employment, Restructuring and Recession*, edited by K. Purcell, S. Wood, A. Watson and S. Allen. London: Macmillan.

Bradley, K. and Gelb, A. 1983. *Cooperation at Work: The Mondragon Experience.* Aldershot: Ashgate.

Braverman, H. 1974. *Labor and Monopoly Capital: The Degradation of Work in the Twentieth Century.* New York: Monthly Review Press.

Brighton Labour Process Group 1977. The capitalist labour process. *Capital and Class*, 1(Spring), 3-26.

Brindley, T., Rydin, Y. and Stoker, G. 1996. *Remaking Planning: Politics of Urban Change in the Thatcher Years.* London: Routledge.

Brummer, A. 2009. *The Crunch: How Greed and Incompetence Sparked the Credit Crisis.* London: Random House Business Books.

Brysk, A. and Shafir, G. (eds) 2004. *People Out of Place: Globalization, Human Rights and the Citizenship Gap.* London: Routledge.

Burawoy, M. 1979. *The Manufacture of Consent Changes in the Labor Process Under Monopoly Capitalism.* Chicago: University of Chicago Press.

Burgoyne, J. 1989. Creating the managerial portfolio: Building on competency approaches to management. *Management Education and Development*, 20(1), 56-61.

Burgoyne, J., Hirsh, W. and Williams, S. 2004. *The Development of Management and Leadership Capability and its Contribution to Performance: The Evidence, the Prospects and the Research Need.* Nottingham: Department for Education and Skills.

Burnham, J. 1941. *The Managerial Revolution.* New York: John Day.

Burrell, G. 1980. Radical organisation theory, in *International Yearbook of Organisation Studies 1979*, edited by D. Dunkerley and G. Salaman. London: Routledge and Kegan Paul, 90-107.

Burrell, G. and Morgan, G. 1979. *Sociological Paradigms and Organisational Analysis.* London: Heinemann.

Cairncross, F. 2001. *The Death of Distance: How the Communications Revolution is Changing our Lives.* Boston: Harvard Business School Press.

Capital and Class Editorial Collective 1982. A socialist GLC in capitalist Britain? *Capital and Class*, 18(Winter), 117-133.

Carchedi, G. 1983. Class analysis and the study of social forms, in *Beyond Method: Strategies for Social Research*, edited by G. Morgan. London: Sage Publications, 347-366.

Carroll, S. and Gillen, D. 1987. Are the classical management functions useful in describing managerial work? *Academy of Management Review*, 12, 38-51.

Castells, M. 1983. *The City and the Grassroots*. London: Edward Arnold.

Castells, M. 2001. *The Internet Galaxy*. Oxford: Blackwell.

Cave, E. and McKeown, P. 1993. Managerial effectiveness: the identification of need. *Management Education and Development*, 2(3), 122-137.

Central Advisory Council for Education 1967. *Children and their Primary Schools (The Plowden Report)*. London: HMSO.

Central Policy Review Staff 1980. *Education, Training and Industrial Performance*. London: Central Policy Review Staff.

Child, J. 1968. British management thought as a case study within the sociology of knowledge. *Sociological Review*, 16(2), 217-239.

Child, J. 1969. *British Management Thought: A Critical Analysis*. London: George Allen and Unwin.

Chitty, C. 1986. TVEI: the MSC's Trojan Horse, in *Challenging the MSC on Jobs, Education and Training*, edited by C. Benn and J. Fairley. London: Pluto Press, 76-98.

Chitty, C. 1989. *Towards a New Educational System: The Victory of the New Right?* Lewes: Falmer Press.

Clawson, D. 1980a. *Bureaucracy and the Labor Process: The Transformation of US Industry, 1860-1920*. New York: Monthly Review Press.

Clawson, D. 1980b. Class struggle and the rise of bureaucracy, in *The International Yearbook of Organization Studies*, edited by D. Dunkerley and G. Salaman. London: Routledge and Kegan Paul, 1-17.

Clegg, S. and Dunkerley, D. (eds) 1978. *Critical Issues in Organizations*. London: Routledge and Kegan Paul.

Clegg, S. and Dunkerley, D. 1980. *Organization, Class and Control*. London: Routledge and Kegan Paul.

Clegg, S. and Hardy, C. 1999. Introduction, in *Studying Organization: Theory and Method*, edited by S. Clegg and C. Hardy. London: Sage Publications, 1-22.

Co-operatives UK. *Co-operative Case Studies* [Online]. Manchester. Available: http://www.cooperatives-uk.coop/Home/miniwebs/miniwebsA-z/caseStudies [accessed 12 January 2010].

Coates, K. and Topham, T. (eds) 1968. *Industrial Democracy in Great Britain: A Book of Readings and Witnesses for Workers' Control*. London: MacGibbon & Kee.

Cockburn, C. 1977. *The Local State: Management of Cities and People*. London: Pluto Press.

Cohen, M. and McBride, S. (eds) 2003. *Global Turbulence.* Aldershot: Ashgate Publishing.

Community Development Project 1977. *Gilding the Ghetto.* London: CDP Inter-Project Editorial Team.

Cooper, C. 2008. *Extraordinary Circumstances: The Journey of a Corporate Whistleblower*. Hoboken: John Wiley and Sons.

Council for Excellence in Management and Leadership 2002. *Managers and Leaders: Raising Our Game*. London: Council for Excellence in Management and Leadership.

Crainer, M. 1999. *The Management Century*. Chichester: Jossey Bass Wiley.

Crozier, M. 1964. *The Bureaucratic Phenomenon.* Chicago: University of Chicago Press.

Curry, J. 1993. The flexible fetish. *Capital and Class*, 50, 99-126.

Czarniawska-Joerges, B. 1998. *A Narrative Approach to Organization Studies*. Thousand Oaks: Sage Publications.

Dahl, R. 1961. *Who Governs: Democracy and Power in an American City*. New Haven: Yale University Press.

Dahrendorf, R. 1957. *Class and Class Conflict in Industrial Society*. Palo Alto: Stanford University Press.

Dar, S. and Cooke, B. (eds) 2008. *The New Development Management.* London: Zed Books.

Datta, R. 2003. From development to empowerment: the Self-Employed Women's Association in India. *International Journal of Politics, Culture, and Society*, 16(3), 351-368.

De Vroey, M. 1975. The separation of ownership and control in large corporations. *Review of Radical Political Economics*, 7(2), 1-10.

Denning, S. 2000. *The Springboard: How Storytelling Ignites Action in Knowledge-Era Organizations*. Woburn: Butterworth-Heinemann.

Derber, C. 2002. *People Before Profit: The New Globalization in an Age of Terror, Big Money, and Economic Crisis*. New York: St Martin's Press.

Devine, F., Savage, M., Scott, J. and Crompton, R. (eds) 2005. *Rethinking Class: Culture, Identities and Lifestyle*. Basingstoke: Palgrave Macmillan.

Dowla, A. and Barua, D. 2006. *The Poor Always Pay Back: The Grameen II Story.* Herndon: Kumarian.

Drucker, P. 1968. *The Practice of Management*. London: Pan.

Edwards, R. 1979. *The Contested Terrain: The Transformation of the Workplace in the Twentieth Century*. London: Heinemann.

Effrat, A. 1973. Power to the paradigm: an editorial introduction. *Sociological Inquiry*, 42(3/4), 3-33.

Ehrenreich, B. and Ehrenreich, J. 1979. The professional-managerial class, in *Between Labour and Capital*, edited by P. Walker. Hassocks: The Harvester Press, 5-45.

Eichenwald, K. 2005. *Conspiracy of Fools*. New York: Broadway Books.

Eisenstadt, S.N. and Cureleu, M. 1976. *The Form of Sociology: Paradigms and Crises*. New York: Wiley.

Elkind, P. and McLean, B. 2004. *The Smartest Guys in the Room: The Amazing Rise and Scandalous Fall of Enron*. London: Penguin.

Elliott, L., Schroth, R. and Elliot, A. 2002. *How Companies Lie: Why Enron is Just the Tip of the Iceberg*. New York: Crown Business.

Engels, F. 1955. *Anti-Duhring*. London: Lawrence and Wishart.

Fayol, H. 1949. *General and Industrial Administration*. London: Sir Isaac Pitman and Sons.

Flew, A. (ed.) 1979. *A Dictionary of Philosophy*. London: Macmillan.

Freeman, J. 1972. The tyranny of structurelessness. *Berkeley Journal of Sociology*, 17, 151-164.

Friedman, A. 1977a. *Industry and Labour: Class Struggle at Work and Monopoly Capitalism*. London: Macmillan.

Friedman, A. 1977b. Responsible autonomy versus direct control over the labour process. *Capital and Class*, 1(Spring), 43-57.

Friedrichs, R. 1970. *A Sociology of Sociology*. New York: Free Press.

Frith, S. 1980. Education, training and the labour process, in *Blind Alley: Youth in a Crisis of Capital*, edited by M. Cole and B. Skelton. Ormskirk: G.W. and A. Hesketh, 25-44.

Gabriel, Y. 2000. *Storytelling in Organizations: Facts, Fictions, and Fantasies*. Oxford: Oxford University Press.

Geddes, M. 1988. The capitalist state and the local economy: 'Restructuring for labour' and beyond. *Capital and Class*, 12(2), 85-120.

Geertz, C. 1973. *The Interpretation of Cultures: Selected Essays*. New York: Basic Books.

Giddens, A. 1984. *The Constitution of Society*. Cambridge: Polity Press.

Gilbreth, F. 1912. *Primer of Scientific Management*. New York: Van Nostrand Rheinhold.

Gillborn, D. 2005. Education policy as an act of white supremacy: whiteness, critical race theory and education reform. *Journal of Education Policy*, 20(4), 485-505.

Gioia, D. and Pitre, E. 1990. Multiparadigm perspectives on theory building. *Academy of Management Review*, 15(4), 584-602.

Gorz, A. 1970. The role of immigrant labour. *New Left Review*, 61(May-June), 28-31.

Greater London Council 1983. *The GLC's Work to Assist Ethnic Minorities*. London: Greater London Council.

Greater London Council 1984. *GLC Grant Aid to Voluntary and Community Organisations*. London: Greater London Council.

Greater London Council 1985. *London Industrial Strategy*. London: Greater London Council.

Greater London Council 1986. *The London Labour Plan*. London: Greater London Council.

Greater London Council Intelligence Unit 1986. *London Facts and Figures*. London: Greater London Council.

Greater London Training Board 1983. *The Youth Training Scheme in London*. London: Greater London Training Board.

Greater London Training Board 1984a. *Review of the New Training Initiative 1981-4*. Greater London Training Board.

Greater London Training Board 1984b. *Training in Crisis Conference Report*. London: Greater London Training Board.

Greater London Training Board 1986. *London Labour Plan*. London: Greater London Training Board.

Green, M., Grieco, M. and Holmes, L. 2002. Archiving social practice: The management of transport boycotts, in *Organising in the Information Age: Distributed Technology, Distributed Leadership, Distributed Identity, Distributed Discourse*, edited by L. Holmes, D. Hosking and M. Grieco. Aldershot: Ashgate, 80-93.

Greene, A. 2001. *Voices from the Shopfloor: Dramas of the Employment Relationship*. Aldershot: Ashgate Publishing.

Gregg, P. 1990. The evolution of special employment measures. *National Institute Economic Review*, 132(1), 49-58.

Grey, C. and Willmott, H. (eds) 2005. *Critical Management Studies: A Reader*. Oxford: Oxford University Press.

Grieco, M. 1998. *Meeting the Moment: Microfinance and the Social Exclusion Agenda* [Online]. London: paper presented to Accounting SIG, Business School, University of North London, 9 December 1998. Available: http://www.transportandsociety.com/changepage/microcredit.htm [accessed 15 January 2010].

Grieco, M. and Whipp, R. 1985. Women and control in the workplace: Gender and control in the workplace, in *Job Redesign: Critical Perspectives on the Labour Process*, edited by D. Knights and H. Willmott. Aldershot: Gower.

Hales, C. 1980. *Managerial Effectiveness and its Determinants: Golden Fleece or Chimera?* London: Training Services Division, Manpower Services Commission.

Hales, C. 2001. *Managing Through Organization: The Management Process, Forms of Organization and the Work of Managers*. 3rd Edition. London: Business Press.

Hamlyn, D. 1953. Behaviour. *Philosophy*, 28(April), 132-145.

Harré, R. and Secord, P. 1972. *The Explanation of Social Behaviour*. Oxford: Blackwell.

Harvey, L. 1982. The use and abuse of Kuhnian Paradigms in the sociology of knowledge. *Sociology*, 16(1), 85-101.

Hassard, J. 1990. An alternative to paradigm incommensurability, in *The Theory and Philosophy of Organizations*, edited by J. Hassard and D. Pym. London: Routledge, 219-230.

Hassard, J. 1991. Multiple paradigms and organizational analysis: A case study. *Organization Studies*, 12(2), 275-299.

Hassard, J. and Pym, D. (eds) 1990. *The Theory and Philosophy of Organizations: Critical Issues and New Perspectives.* London: Routledge.

Heydebrand, W.V. 1983. Organisation and praxis, in *Beyond Method: Strategies for Social Research*, edited by G. Morgan. London: Sage Publications, 306-320.

Hodgson, G. 2002. The legal nature of the firm and the myth of the firm-market hybrid. *International Journal of the Economics of Business*, 9(1), 37-60.

Holmes, L. 2003. Reaching in, reaching out: metadata, popular planning and social capital development. *European Spatial Research and Policy*, 10(2), 25-38.

Holmes, L. and Grieco, M. 1991. Overt funding, buried goals, and moral turnover: the organizational transformation of radical experiments. *Human Relations*, 44(7), 643-664.

Holmes, L. and Grieco, M. 2001. The power of transparency: the Internet, email, and the Malaysian political crisis. *Asia-Pacific Business Review*, 8(2), 59-72.

Hosking, D., Dachler, H. and Gergen, K. 1995. *Management and Organization: Relational Alternatives to Individualism*. Aldershot: Avebury.

HPCC Bus Garage Project Steering Group 1984. *From the Stonebridge Bus Garage ... to the Stonebridge Community Complex*. London: HPCC Bus Garage Project Steering Group.

Jackson, N. and Carter, P. 1991. In defence of paradigm incommensurability. *Organization Studies*, 12(1), 109-127.

Jenkins, G. and Poole, M. (eds) 1990. *New Forms of Ownership: Management and Employment.* London: Routledge.

Jeter, L. 2003. *Disconnected: Deceit and Betrayal at WorldCom*. Hoboken: John Wiley and Sons.

Kamoche, J. 2000. *Sociological Paradigms and Human Resources: An African Context*. Aldershot: Ashgate.

Kettle, M. and Hodges, L. 1982. *Uprising! The Police, the People and the Riots in Britain's Cities*. London: Pan.

Kivisto, P. and Hartung, E. (eds) 2006. *Intersecting Inequalities: Class, Race, Sex and Sexualities*. Harlow: Pearson Education.

Klein, N. 2000. *No Logo*. London: Flamingo.

Knights, D. and Willmott, H. (eds) 1989. *Labour Process Theory*. London: Macmillan.

Kotter, J. 1982. *The General Managers*. New York: Free Press.

Kuhn, T. 1970. *The Structure of Scientific Revolutions*, 2nd Edition. Chicago: University of Chicago Press.

Lee, D. and Turner, B. (eds) 1996. *Conflicts about Class: Debating Inequality in Late Industrialism.* Harlow: Longman.

Lees, R. and Mayo, M. 1984. *Community Action for Change*. London: Routledge and Kegan Paul.

Levitas, R. and Guy, W. 1996. *Interpreting Official Statistics*. London: Routledge.

Lindenfield, F. and Rothschild-Whitt, J. (eds) 1982. *Workplace Democracy and Social Change*. Boston: Porter Sargent Publishing.

Little, S., Holmes, L. and Grieco, M. 2001. Island histories, open cultures?: The electronic transformation of adjacency. *South African Business Review*, 4(2), 21-25.

Littler, C. 1982. *The Development of the Labour Process in Capitalist Societies*. London: Heinemann.

Loney, M. 1983. *Community Against Government: British Community Development Project, 1968-1978*. London: Heinemann.

Lupton, T. 1983. *Management and the Social Sciences*, 3rd Edition. Harmondsworth: Penguin.

Mander, J. and Tauli-Corpuz, V. 2006. *Paradigm Wars: Indigenous Peoples' Resistance to Globalization*. San Francisco: Sierra Club Books.

Manpower Services Commission 1980. *Outlook on Training*. London: Manpower Services Commission.

Manpower Services Commission 1981a. *A Framework for the Future: A Sector by Sector Review of Industrial and Commercial Training*. London: Manpower Services Commission.

Manpower Services Commission 1981b. *A New Training Initiative: A Consultative Document*. London: Manpower Services Commission.

Manpower Services Commission 1982a. *Community Programme: The Agents' and Sponsors' Handbook*. London: Manpower Services Commission.

Manpower Services Commission 1982b. *Community Programme: Information for Sponsors*. London: Manpower Services Commission.

Manpower Services Commission 1983. *Towards an Adult Training Strategy: A Discussion Paper*. London: Manpower Services Commission.

Mant, A. 1979. *The Rise and Fall of the British Manager*. Revised Edition. London: Pan Books.

Mantle, A. 1985. *Popular Planning NOT in Practice*. London: Greenwich Economic Resource Unit.

Marsh, S. 1986. Women and the MSC, in *Challenging the MSC on Jobs, Education and Training*, edited by C. Benn and J. Fairley. London: Pluto Press, 153-176.

Marx, K. 1953a. The Eighteenth Brumaire of Louis Bonaparte, in *Karl Marx and Frederick Engels: Selected Works, vol. I*, edited by K. Marx and F. Engels. London: Lawrence and Wishart, 221-311.

Marx, K. 1953b. Preface to a contribution to the critique of political economy, in *Karl Marx and Frederick Engels: Selected Works, vol. I*, edited by K. Marx and F. Engels. London: Lawrence and Wishart, 327-331.

Marx, K. 1954. *Capital*. London: Lawrence and Wishart.

Marx, K. and Engels, F. 1953. Manifesto of the Communist Party, in *Karl Marx and Frederick Engels: Selected Works, Vol. I*, edited by K. Marx and F. Engels. London: Lawrence and Wishart, 21-61.

Matthäus-Maier, I. 2009. *New Partnerships for Innovation in Microfinance*. Berlin: Springer-Verlag.

Maud, J. 1967. *Report of the Committee on Management of Local Government*. London: HMSO.

Maynard, M. (ed.) 1994. *The Dynamics of Race and Gender*. London: Taylor and Francis.

Messinger, S. 1955. Organizational transformation: A case study of a declining social movement. *American Sociological Review*, 20(1), 3-10.

Miliband, R. 1969. *The State in Capitalist Society*. London: Weidenfield & Nicolson, reprinted 1973 by Quartet.

Mills, C. 1956. *The Power Elite*. New York: Oxford University Press.

Mintzberg, H. 1973. *The Nature of Managerial Work*. New York: Harper and Row.

Moore, R. 1975. *Racism and Black Resistance*. London: Pluto Press.

Moore, R. 1983. Further education, pedagogy and production, in *Youth Training and the Search for Work*, edited by D. Gleeson. London: Routledge and Kegan Paul, 14-31.

National Association of Teachers in Further and Higher Education and Association for Adult and Continuing Education 1983. *Adult Unemployment: A Discussion Paper*. National Association of Teachers in Further and Higher Education/ Association for Adult and Continuing Education.

National Council for Voluntary Organisations 1985. *The Urban Programme Explained, Information Sheet No 27*. National Association of Teachers in Further and Higher Education/Association for Adult and Continuing Education.

Nelson, N. and Wright, S. (eds) 1997. *Power and Participatory Development*. London: Intermediate Technology Publications.

Network Training Group 1983. *Training and the State: Responses to the Manpower Services Commission*. Leeds: Network Training Group.

Nichols, T. 1969. *Ownership, Control and Ideology*. London: George Allen and Unwin.

Norman, D. 1988. *The Design of Everyday Things*. New York: Currency-Doubleday.

Office for National Statistics 2000. *Standard Occupational Classification 2000, Volume 1: Structure and Descriptions of Unit Groups*. London: The Stationery Office.

Office for National Statistics 2006. *Labour Market Trends*. London: Palgrave Macmillan.

Organisation for Economic Co-operation and Development 2003. *Asset Building and the Escape from Poverty: A New Welfare Policy Debate*. Paris: Organisation for Economic Co-operation and Development.

Parker, M., Fournier, V. and Reedy, P. 2007. *The Dictionary of Alternatives: Utopianism and Organization*. London: Zed Books.

Pearson, R. 1986. Female workers in the first and third worlds: The 'greening' of women's labour, in *The Changing Experience of Employment, Restructuring and Recession*, edited by K. Purcell, S. Wood, A. Watson and S. Allen. London: Macmillan.

Phizacklea, A. and Miles, R. 1980. *Labour and Racism*. London: Routledge & Kegan Paul.
Pickerill, J. 2006. Radical politics on the net. *Parliamentary Affairs*, 59(2), 266-282.
Polet, F. (ed.) 2004. *Globalizing Resistance: The State of the Struggle*. London: Pluto Press.
Pollard, H. 1974. *Developments in Management Thought*. London: Heinemann.
Pollard, S. 1965. *The Genesis of Modern Management*. London: Edward Arnold.
Pollert, A. 1988. The 'flexible firm': Fixation or fact?, *Work, Employment and Society*, 2(3), 281-316.
Pollert, A. (ed.) 1991. *Farewell to Flexibility*. Oxford: Blackwell.
Polsby, N. 1963. *Community Power and Political Theory*. New Haven: Yale University Press.
Poulantzas, N. 1969. The problem of the capitalist state. *New Left Review*, 58(November/December), 67-78.
Poulantzas, N. 1973. On social classes. *New Left Review*, 78(March-April), 27-54.
Poulantzas, N. 1975. *Classes in Contemporary Capitalism*. London: NLB.
Presthus, R. 1979. *The Organizational Society*. Basingstoke: Macmillan.
Price, R. and Bain, G. 1988. The labour force, in *British Social Trends Since 1900*, edited by A. Halsey. Basingstoke: Macmillan.
Rampton, A. 1981. *West Indian Children in our Schools: Interim Report of the Committee of Inquiry into the Education of Children from Ethnic Minority Groups*. London: HMSO.
Reed, M. 1999. Organizational theorizing: A historically contested terrain, in *Studying Organization: Theory and Method*, edited by S. Clegg and C. Hardy. London: Sage Publications, 25-50.
Robbins, S. and Coulter, M. 2005. *Management*, 8th Edition. Upper Saddle River: Pearson Education.
Roethlisberger, F. and Dickson, W. 1939. *Management and the Worker*. Cambridge: Harvard University Press.
Rose, K. 1992. *Where Women are Leaders: The SEWA Movement in India*. London: Zed Books.
Ryan, W. 1976. *Blaming the Victim*. New York: Vintage Books.
Ryle, G. 1949. *The Concept of Mind*. London: Hutchinson.
Salaman, G. 1979. *Work Organisations: Resistance and Control*. London: Longman.
Sarup, M. 1982. *Education, State and Crisis: A Marxist Perspective*. London: Routledge & Kegan Paul.
Scarman, L. 1982. *The Scarman Report: The Brixton Disorders 10-12 April 1981*. Harmondsworth: Penguin.
Scofield, P., Preston, P. and Jacques, E. 1983. *Youth Training: The Tories' Poisoned Apple*. London: Independent Labour Publications.
Sheldrake, J. 1996. *Management Theory: from Taylorism to Japanization*. London: International Thomson Business Press.

Sherman, H. 1976. Dialectics as a method. *Insurgent Sociologist*, 6(4), 57-64.
Sherman, H. and Wood, J. 1979. *Sociology: Traditional and Radical Perspectives*. New York: Harper and Row.
Sills, D. 1957. *The Volunteers: Means and Ends in a National Organization*. Glencoe: The Free Press.
Simon, B. 1991. *Education and the Social Order 1940-1990*. London: Lawrence and Wishart.
Sivandan, A. 1976. Race, class and the state: The Black experience in Britain. *Race and Class*, 17(4).
Sivandan, A. 1982. *A Different Hunger: Writings on Black Resistance*. London: Pluto Press.
Smith, G. 1987. Whatever happened to Educational Priority Areas? *Oxford Review of Education*, 13(1), 23-38.
Society of Civil and Public Servants 1982. *Back to Work: An Alternative Strategy for The Manpower Services Commission*. London: Society of Civil and Public Servants.
St. John-Brooks, C. 1985. *Who Controls Training?: The Rise of the Manpower Services Commission*. London: Fabian Society.
Stewart, M. and Whitting, G. 1983. *Ethnic Minorities and the Urban Programme*. Bristol: School for Advanced Urban Studies.
Stewart, R. 1967. *The Reality of Management*. London: Pan.
Stewart, R. 1976. *Contrasts in Management*. London: Pan.
Stone, K. 1974. The origins of job structures in the steel industry. *Review of Radical Political Economic*, 6(2), 113-173.
Stonebridge Bus Depot Steering Group 1981. *Stonebridge Bus Depot Project Report*. London: Stonebridge Bus Depot Steering Group.
Storey, J. 1982. *Managerial Prerogative and the Question of Control*. London: Routledge and Kegan Paul.
Swingewood, A. 1975. *Marx and Modern Social Theory*. Basingstoke: Macmillan.
Taylor, F. 1903. *Shop Management*. New York: Harper and Row.
Therborn, G. 1970. What does the ruling class do when it rules? *Insurgent Sociologist*, 6(3), 1-16.
Therborn, G. 1978. *What Does the Ruling Class do When it Rules?* London: NLB.
Therborn, G. 1982. What does the ruling class do when it rules?, in *Class, Power and Conflict: Classical and Contemporary Debates*, edited by A. Giddens and D. Held. London: Macmillan, 224-248.
Thompson, P. 1983. *The Nature of Work*. London: Macmillan.
Thompson, P. and McHugh, D. 2002. *Work Organisations: A Critical Introduction*. Basingstoke: Palgrave.
Ure, A. 1835. *The Philosophy of Manufactures*. London: Charles Knight.
Urwick, L. 1943. *The Elements of Administration*. London: Pitman.
Urwick, L. and Brech, E. 1947. *The Making of Scientific Management, Volume III: The Hawthorne Experiments*. London: Pitman.

Walker, B. 1988. From dole college to Youth Training Scheme: the development of state intervention in 'training' unemployed youth. *Journal of Further and Higher Education*, 12(1), 20-41.

Walker, P. (ed.) 1979. *Between Labour and Capital.* Brighton: Harvester Press.

Wallace, C., Joshua, H. and Booth, H. 1981. *To Ride The Storm: The 1981 Bristol Rioting and the State.* London: Heinemann.

Ward, M. 1983. Labour's capital gains: Greater London Council experience. *Marxism Today*, December, 24-29.

Ward, M. 1985. Making the organisation work for members. *Local Government Studies*, 11(4).

Weaver, G. and Gioia, D. 1994. Paradigms lost: Incommensurability vs structurationist inquiry. *Organization Studies*, 15(4), 565-590.

Weber, L. 2001. *Understanding Race, Class, Gender, and Sexuality: A Conceptual Framework.* Boston: McGraw-Hill.

Webster, F. 2002. *Theories of the Information Society.* 2nd Edition. London: Routledge.

Weihrich, H. and Koontz, H. 1993. *Management: A Global Perspective.* 10th International Edition. New York: McGraw-Hill.

Wellman, B. and Haythornthwaite, C. (eds) 2002. *The Internet in Everyday Life.* Malden: Blackwell.

Wickham, G. 1985. Gender divisions, training and the state, in *Education, Training and Employment*, edited by R. Dale. Oxford: Pergamon, 95-110.

Wilhelm, A. 2000. *Democracy in the Digital Age: Challenges to Political Life in Cyberspace.* New York: Routledge.

Williams, P. 1992. *'It Can Be Done': The Story Behind Bridge Park.* London: Ronald Mann.

Williams, R. 1976. *Keywords.* London: Fontana.

Williams, S. 2001. *Diversity in the Management and Leadership Population.* London: Council for Excellence in Management and Leadership.

Williamson, O. 1975. *Markets and Hierarchies.* New York: The Free Press.

Willmott, H. 1993. Breaking the paradigm mentality. *Organization Studies*, 14(5), 681-719.

Wittgenstein, L. 1953. *Philosophical Investigations.* Oxford: Blackwell.

Wood, S. (ed.) 1982. *The Degradation of Work? Skill, Deskilling and the Labour Process.* London: Hutchinson.

Wright, E. 1985. *Classes.* London: Verso.

Wright, E.O. 1983. *Class, Crisis and the State.* London: Verso.

Young, M. and Rigge, M. 1983. *Revolution from Within.* London: George Weidenfeld and Nicolson.

Yunus, M. and Jolis, A. 1998. *Banker to the Poor.* London: Aurum Press Limited.

Zald, M. and Denton, P. 1963. From evangelism to general service: The transformation of the YMCA. *Administrative Science Quarterly*, 8, 218-234.

Zey-Ferrell, M. and Aitken, M. 1981. *Complex Organisations: Critical Perspectives.* Glenview: Scott, Foresman and Company.

Index